T0258412

Applied Principles of Operations Management

Applied Principles of Operations Management

Edited by **Edward Pepper**

New York

Published by NY Research Press,
23 West, 55th Street, Suite 816,
New York, NY 10019, USA
www.nyresearchpress.com

Applied Principles of Operations Management
Edited by Edward Pepper

International Standard Book Number: 978-1-63238-057-9 (Hardback)

Printed in the United States of America.

Contents

Preface

This book has been a concerted effort by a group of academicians, researchers and scientists, who have contributed their research works for the realization of the book. This book has materialized in the wake of emerging advancements and innovations in this field. Therefore, the need of the hour was to compile all the required researches and disseminate the knowledge to a broad spectrum of people comprising of students, researchers and specialists of the field.

The applied principles of operations management have been elucidated in this up-to-date book. It illustrates the important connection between operations strategy, its management and research, and various departments, systems and applications throughout an organization. The contributors within the book depict how mathematical devices and enhanced procedures can be applied effectively in unique measures to other operations. It gives examples that demonstrate the challenges confronting the firms competing in today's demanding environment in overcoming the gap between theory and practice by evaluating real situations.

At the end of the preface, I would like to thank the authors for their brilliant chapters and the publisher for guiding us all-through the making of the book till its final stage. Also, I would like to thank my family for providing the support and encouragement throughout my academic career and research projects.

Editor

Part 1

Competing with Operations

Lean Six Sigma in the Service Industry

Alessandro Laureani
University of Strathclyde
United Kingdom

1. Introduction

The business improvement methodology known as Lean Six Sigma is rooted in the manufacturing industry, where it developed over the past few decades, reaching widespread adoption worldwide. However, according to the *World Economic Outlook Database*, published in April 2011, by the International Monetary Fund (IMF, 2011), the distribution of PPP (Purchase Power Parity) GDP, in 2010, among various industry sectors in the main worldwide economies, reflected a decline in the industrial sector, with the service sector now representing three-quarters of the US economy and more than half of the European economies.

PPP GDP 2010	Agriculture	Industry	Service
European Union	5.7%	30.7%	63.6%
United States	1.2%	22.2%	76.7%
China	9.6%	46.8%	43.6%
India	16.1%	28.6%	55.3%

Table 1. PPP GDP Sector Comparison 2010.

In light of the increasing importance of the service sector, the objective of this chapter is to discuss whether the business improvement methodology known as Lean Six Sigma is applicable to the service industry as well, and illustrate some case study applications.

2. What is Lean Six Sigma?

Lean Six Sigma is a business improvement methodology that aims to maximize shareholders' value by improving quality, speed, customer satisfaction, and costs. It achieves this by merging tools and principles from both Lean and Six Sigma. It has been widely adopted widely in manufacturing and service industries, and its success in some famous organizations (e.g. GE and Motorola) has created a copycat phenomenon, with many organizations across the world willing to replicate the success.

Lean and Six Sigma have followed independent paths since the 1980s, when the terms were first hard-coded and defined. The first applications of Lean were recorded in the Michigan plants of Ford in 1913, and were then developed to perfection in Japan (within the Toyota Production System), while Six Sigma saw the light in the United States (within the Motorola Research Centre).

Lean is a process-improvement methodology, used to deliver products and services better, faster, and at a lower cost. Womack and Jones (1996) defined it as:

… a way to specify value, line up value-creating actions in the best sequence, conduct those activities without interruption whenever someone requests them, and perform them more and more effectively. In short, lean thinking is lean because it provides a way to do more and more with less and less — less human effort, less human equipment, less time, and less space — while coming closer and closer to providing customers with exactly what they want. (Womack and Jones, 1996:p.)

Six Sigma is a data-driven process improvement methodology used to achieve stable and predictable process results, reducing process variation and defects. Snee (1999) defined it as: 'a business strategy that seeks to identify and eliminate causes of errors or defects or failures in business processes by focusing on outputs that are critical to customers'.

While both Lean and Six Sigma have been used for many years, they were not integrated until the late 1990s and early 2000s (George, 2002; George, 2003). Today, Lean Six Sigma is recognized as: 'a business strategy and methodology that increases process performance resulting in enhanced customer satisfaction and improved bottom line results' (Snee, 2010).

Lean Six Sigma uses tools from both toolboxes, in order to get the best from the two methodologies, increasing speed while also increasing accuracy.

The benefits of Lean Six Sigma in the industrial world (both in manufacturing and services) have been highlighted extensively in the literature and include the following:

1. Ensuring services/products conform to what the customer needs ('voice of the customer').
2. Removing non-value adding steps (waste) in critical business processes.
3. Reducing the cost of poor quality.
4. Reducing the incidence of defective products/transactions.
5. Shortening the cycle time.
6. Delivering the correct product/service at the right time in the right place. (Antony, 2005a; Antony, 2005b)

Examples of real benefits in various sectors are illustrated in Table 2.

One of the key aspects differentiating Lean Six Sigma from previous quality initiatives is the organization and structure of the quality implementation functions. In quality initiatives prior to Lean Six Sigma, the management of quality was relegated largely to the production floor and/or, in larger organizations, to some statisticians in the quality department. Instead, Lean Six Sigma introduces a formal organizational infrastructure for different quality implementation roles, borrowing terminology from the world of martial arts to define hierarchy and career paths (Snee, 2004; Antony, Kumar & Madu, 2005c; Antony, Kumar & Tiwarid, 2005d; Pande, Neuman & Cavanagh, 2000; Harry & Schroeder, 1999; Adams, Gupta & Wilson, 2003).

Service	Problem	Outcome	Benefits
Healthcare	Increase radiology throughput and decrease cost per radiology procedure in a hospital (Thomerson, 2001)	Significant improvement in radiology throughput and reduction in cost per radiology procedure	33 per cent increase in radiology throughput
	Poor patient safety due to high medication and laboratory errors (Buck, 2001)	Reduced medication and laboratory errors	22 per cent reduction in cost per radiology procedure $1.2 million in savings
	Overcrowded emergency department (Revere and Black, 2003)	Reduced time to transfer a patient from the ER to an inpatient hospital bed	Improved patient safety significantly $600,000/year in profit
Banking	Reduce customer complaints (Roberts, 2004)	Significant reduction in customer complaints and increase in customer satisfaction	10.4 per cent increase in customer satisfaction 24 per cent decrease in customer complaints
	Excessive internal and external call backs plus unacceptable credit processing time (Rucker, 2000)	Reduction in both internal and external call backs, reduction in credit processing time	Reduced internal call backs by 80 per cent
	High number of flaws in customer-facing processes (e.g. account opening, payment handling, etc.) (www.helpingmakingithappen.com)	Reduced flaws in all customer-facing processes	Increased customer satisfaction Improved process efficiency Reduced cycle time by over 30 per cent
	High returned renewal credit cards per month in a leading bank (Keim, 2001)	Significant reduction in the number of returned renewal credit cards	Defect rate reduced from 13,500 DPMO to 6,000 DPMO
	Excessive market losses on trading errors, high costs associated with electronic order corrections etc. in an investment banking unit (Stusnick, 2005)	Reduced trading errors significantly Reduced costs associated with order corrections, etc.	Several millions of dollars in savings Improved employee morale within the banking unit
Financial services	High administrative costs (www.executiveonline.co.uk)	Reduction in administration costs	Savings generated from this project are approximately $75,000/year
	Unacceptable wire transfer processing time to customers	Reduced wire transfer processing time by 40 per cent	Savings generated from the project are around $700,000/year
	Problems in accounts receivables within an accounting department (www.ssqi.com)	Improved cash flow	Annual savings are estimated to be well over $350,000
Utility services	Poor service delivery (www.executiveonline.co.uk)	Improved service delivery	Annual savings from the project is of the order of over $1.5 million
	High contract complaints resulted in customer dissatisfaction and high costs	Reduced the number of complaints after six sigma methodology was introduced	Complaints reduced from 109 to 55 on average per year
Miscellaneous	Poor delivery performance in a logistics company (Thawani, 2004)	Reduced the number of delayed deliveries	Sigma quality level of the process improved from 2.43 (176,000 DPMO) to 3.94 (7,400 DPMO) Improved customer satisfaction and increased market share, resulted in savings of $400,000 (approx.)
	Significant errors in a monthly publication for Wall Street investors and traders	Reduction in reporting and accounting errors	$1.2 million in estimated savings

Table 2. Benefits of Six Sigma in Service Organizations (Antony, Kumar & Cho, 2007).

3. Lean Six Sigma and the service industry

The service industry has its own special characteristics, which differentiate it from manufacturing and make it harder to apply Lean Six Sigma tools, which can be summarized in the following main areas (Kotler, 1997; Regan 1963; Zeithmal, Parasur and Berry 1985):

Intangibility: Although services can be consumed and perceived, they cannot be measured easily and objectively, like manufacturing products. An objective measurement is a critical aspect of Six Sigma, which requires data-driven decisions to eliminate defects and reduce variation. The lack of objective metrics is usually addressed in service organizations through the use of proxy metrics (e.g. customer survey).

Perishability: Services cannot be inventoried, but are instead delivered simultaneously in response to the demand for them. As a consequence, services processes contain far too much 'work-in-process' and work can spend more than 90% of its time waiting to be executed (George, 2003).

Inseparability: Delivery and consumption of service is simultaneous. This adds complexity to service processes, unknown to manufacturing. Having customers waiting in line or on the phone involves some emotional management, not present in a manufacturing process.

Variability: Each service is a unique event dependent on so many changing conditions, which cannot be reproduced exactly. As a result of this, the variability in service processes is much higher than in manufacturing processes, leading to very different customer experiences.

Owing to these inherent differences, it has been harder for service organizations, such as financial companies, health-care providers, retail and hospitality organizations, to apply Lean Six Sigma to their own reality. However, there are also great opportunities in the service organizations (George 2003):

- Empirical data has shown the cost of services are inflated by 30–80% of waste.
- Service functions have little or no history of using data to make decisions. It is often difficult to retrieve data and many key decision-makers may not be as 'numerically literate' as some of their manufacturing counterparts.
- Approximately 30–50% of the cost in a service organization is caused by costs related to slow speed, or carrying out work again to satisfy customer needs.

In the last few years, successful applications in service organizations have come to fruition and we will illustrate three possible applications: in a call centre, in human resources, and finally in a healthcare provider.

4. Case study 1: Lean Six Sigma in a call centre (Laureani et al, 2010a)

The two major types of call centres are outbound centres and inbound centres. The most common are inbound call centre operations. Almost everyone in their daily life has had to call one of those centres for a variety of reasons. Outbound centres are used more in areas such as marketing, sales and credit collection. In these instances, it is the call centre operators who establish contact with the user.

Although there are some differences between outbound and inbound call centres, they each have certain potential benefits and challenges, with regard to the implementation of Lean Six Sigma.

Benefits

Some of the benefits that Lean Six Sigma can deliver in a call centre are (Jacowski, 2008; Gettys, 2009):

1. Streamlining the operations of the call centre: Lean strategy helps in eliminating waste and other non-value added activities from the process.
2. Decreasing the number of lost calls: Six Sigma's root-cause analysis and hypothesis-testing techniques can assist in determining how much time to spend on different type of calls, thus providing a guide to the operators.
3. Better use of resources (both human resources and technology), thus leading to a reduction in the cost of running such centres.
4. Unveiling the 'hidden factory': establishing the root causes of why customers call in the first place can help in uncovering trouble further along the process, providing benefits that go further than the call centre itself, improving customer service and support.
5. Reducing employee turnover: call centres are usually characterized by high employee turnover, owing to the highly stressful work environment. A more streamlined operation would assist in reducing operators' stress, particularly in an inbound centre.

Challenges

Specific challenges of applying Lean Six Sigma in a call centre environment (Piercy & Rich, 2009):

1. The relentless pace of the activity (often 24/7) makes it more difficult for key staff to find the time to become involved in projects and Lean Six Sigma training.
2. The realization of an appropriate measurement system analysis (MSA) (Wheeler & Lyday, 1990) is difficult because of the inherent subjectivity and interpretation of some call types, failing reproducibility tests of different call centre operators.
3. High employee turnover, that normally characterizes call centres, makes it more difficult for the programme to remain in the organization.

Strengths	Weaknesses
• Root cause analysis can determine major reasons for customers' calls, helping to unveil problems further along the value stream map of the company	• Lean Six Sigma deployment requires significant investment in training, that may be difficult from a time perspective in a fast-paced environment such as a call centre
Opportunities	Threats
• Decrease number of lost calls • Reduce waiting time for calls in the queue • Improve employee productivity (i.e. number of calls dealt with by the hour)	• Lack of metrics • Lack of support from process owner • Preconceived ideas

Table 3. SWOT Analysis for the Use of Lean Six Sigma in a Call Centre.

Overall, the opportunities far outweigh the challenges. Call centres nowadays are more than just operations: they are the first, and sometimes a unique, point of contact that a company may have with its customers. Their efficient and effective running, and their timely resolution of customers' queries, all go a long way to establishing the company's brand and image.

Project selection is a critical component of success. Not all projects may be suitable candidates for the application of Lean Six Sigma, and this needs to be kept in mind in assessing the operation of a call centre. Also, different tools and techniques may be more suited to a specific project, depending on the nature and characteristics of the process it is trying to address.

Projects that better lend themselves to Lean Six Sigma share, *inter alia*, the following characteristics:

• The focus of the project is on a process that is either not in statistical control (*unstable*) or outside customer specifications (*incapable*). As already mentioned in the introduction, Six Sigma techniques focus on reducing the variation in a process, making them the ideal tools for tackling an incapable but stable process, whereas Lean tools focus more on the elimination of waste and would be the first port of call for streamlining an unstable process. Priority should be given to unstable processes, using Lean tools to eliminate the waste and simplify the process. Once it has stabilized, more advanced statistical tools from the Six Sigma toolbox, can be used to reduce variation and make the process capable.

- The root reason(s) for this has not been identified yet. It is important to start work on the project with an open mind and without any prejudice. Data and hard facts should guide the project along its path.
- Quantitative metrics of the process are available. A lack of measures and failing to realize a complete measurement system analysis (MSA) (Wheeler & Lyday, 1990) can seriously jeopardize any improvement effort.
- The process owner is supportive and willing to provide data and resources. This is critical for the ongoing success of the project; the process owner's role is discussed in detail in the Control Phase section.

Potential areas of focus for Six Sigma projects in call centres (Gettys, 2009):

- Lost call ratio out of total calls for an inbound call centre;
- Customer waiting/holding times for an inbound call centre;
- First-call resolution;
- Calls back inflating call volumes.

Call centres are increasingly important for many businesses and are struggling consistently with the pressure of delivering a better service at a lower cost. Lean Six Sigma can improve the operation of a call centre through an increase in first-call resolution (that reduces the failure created by failing to answer the query in the first place), a reduction in call centre operator turnover (leveraging on training and experience), and streamlining the underlying processes, eliminating unnecessary operations.

Given the large scale of many call-centre operations, even a relatively small improvement in the sigma value of the process can dramatically reduce the defect rate, increase customer satisfaction and deliver financial benefits to the bottom line (Rosenberg, 2005).

By focusing on eliminating waste, identifying the real value-adding activities and using the DMAIC tools for problem-solving, it is possible to achieve significant improvements in the cost and customer service provided (Swank, 2003).

5. Case study 2: Lean Six Sigma in HR administration (Laureani & Antony, 2010b)

In the late 1980s, when Motórola implemented Six Sigma originally, obtaining astonishing results, the company was then faced with the dilemma of how to reward its employees for these successes (Gupta, 2005). This was the first time Six Sigma and HR practices came into contact, and a more accurate definition of HR practices was needed.

If, in the past, the term HR was related only to administrative functions (e.g. payroll, timekeeping, etc.), the term has increased substantially, in the last few decades, to include the acquisition and application of skills and strategies to maximize the return on investment from an organization's human capital (Milmore et al, 2007).

HR management is the strategic approach to the management of all people that contribute to the achievement of the objectives of the business (Armstrong, 2006). As such it includes, but it is not limited to, personnel administration. In effect it includes all steps where an employee and an organization come into contact, with the potential of adding value to the organization (Ulrich, 1996).

As such, and merging terminology from Lean and HR, we define the following seven points as the Human Capital Value Stream Map:

1. Attraction
2. Selection
3. Orientation (or induction)
4. Reward
5. Development
6. Management
7. Separation

Fig. 1. Human Capital Value Stream Map.

The Human Capital Value Stream Map is a Lean technique that identifies the flow of information or material required in delivering a product or service to a customer (Womack & Jones, 1996). Human capital is the accumulated skills and experience of the human force in an organization (Becker, 1993).

The Human Capital Value Stream Map is the flow of human capital required for an organization to deliver its products or service to customers; the objectives of which are briefly described below:

*   **Attract**: to establish a proper employer's brand that attracts the right calibre of individual.
*   **Select**: to select the best possible candidate for the job.
*   **Orient**: to ensure new employees are properly trained and integrated into the organization.
*   **Reward**: to ensure compensation packages are appropriate and in line with the market.
*   **Develop**: to distinguish talent and ensure career progression.
*   **Manage**: to supervise and administer the day-to-day jobs.
*   **Separation**: to track reasons for voluntary leavers and maintain a constructive relationship.

It is possible to apply Lean Six Sigma tools to each step of the Human Capital Value Stream Map, in order to eliminate waste in the HR process (Wyper & Harrison, 2000). For each step in the Human Capital Value Stream Map it is necessary to establish proper quantitative metrics that allow objective assessment and control of the process step (Sullivan, 2003). This makes use of the more quantitative statistical tools from the Six Sigma toolbox possible.

Establishing HR metrics can be controversial, with different parts of the organization having different objectives (Jamrog & Overholt, 2005), but the answer to these simple questions may help to focus on the real value each step can provide.

1. What is the expected deliverable of the step?
2. What are the relevant metrics and key performance indicators of the step?
3. What are the opportunities for defects in the step?

For recruitment, for example, the answers to the above questions may be as follows.

1. Hire, in the shortest possible time, new members of staff to fulfil a certain job.
2. The number of days to fill a vacancy (also define the acceptable norm for the organization).
3. Any job remaining vacant for longer than the acceptable norm.

Similar thought processes can be performed for other steps: having set metrics for each step of the Human Capital Value Stream Map, an organization is now in the position to apply Six Sigma DMAIC to it.

Six Sigma can be used to improve administrative processes, such as HR processes. Implementing the Six Sigma DMAIC breakthrough methodology in HR follows the same path as implementing it in any other part of the organization.

However, there are some specific key learning points and challenges for the HR area, such as:

• Difficulty in establishing an appropriate measurement system analysis and metrics;
• Data collection can be extremely difficult, as the project team is dealing with very sensitive issues; and
• Difficulty in performing any pilot or design of experiment. Any of these is going to impact on the behaviour of staff, making it difficult to measure its results accurately.

As a result, projects may last longer than the standard four to six months and the wider use of tools such as brainstorming and 'Kaizen' workshops with domain experts may be necessary (Lee et al, 2008).

Examples of potential Six Sigma projects in the HR function are:

• reduction of employees' turnover
• reduction in time and cost to hire a new employee
• reduction in training costs
• reduction in cost of managing employees' separation
• reduction in administrative defects (payroll, benefits, sick pay, etc.)
• reduction in queries from the employee population to the HR department.

Every area of an organization needs to perform better, faster and more cheaply, to keep the company ahead of the competition, and be able to satisfy ever-increasing customer expectations. HR is no exception: more cost-effective and streamlined HR processes will create value for the organization, instead of just being a support act for management (Gupta, 2005).

6. Case study 3: Lean Six Sigma in health-care delivery

Health care is a complex business, having to balance continuously the need for medical care and attention to financial data. It offers pocket of excellence, with outstanding advances in technology and treatment, together with inefficiencies and errors (Taner et al, 2007). Everywhere in the world, the financial pressures on health care have increased steadily in the last decade. While an ageing population and technological investments are often cited as culprits for these financial pressures, unnecessary operational inefficiency is another source of cost increases, largely under the control of health-care professionals (de Koning et al, 2006).

Lean Six Sigma projects so far in the health-care literature have focused on direct care delivery, administrative support and financial administration (Antony et al, 2006), with projects executed in the following processes (Taner et al, 2007):

- increasing capacity in X-ray rooms
- reducing avoidable emergency admissions
- improving day case performance
- improving accuracy of clinical coding
- improving patient satisfaction in Accident and Emergency (A&E)
- reducing turn-around time in preparing medical reports
- reducing bottle necks in emergency departments
- reducing cycle time in various inpatient and outpatient diagnostic areas
- reducing number of medical errors and hence enhancing patient safety
- reducing patient falls
- reducing errors from high-risk medication
- reducing medication ordering and administration errors
- improving active management of personnel costs
- increasing productivity of health-care personnel
- increasing accuracy of laboratory results
- increasing accuracy of billing processes and thereby reducing the number of billing errors
- improving bed availability across various departments in hospitals
- reducing number of postoperative wound infections and related problems
- improving MRI exam scheduling
- reducing lost MRI films
- improving turn-around time for pharmacy orders
- improving nurse or pharmacy technician recruitment
- improving operating theatre throughput
- increasing surgical capacity
- reducing length of stay in A&E
- reducing A&E diversions
- improving revenue cycle
- reducing inventory levels
- improving patient registration accuracy
- improving employee retention

The focus has been on the improvement of clinical processes to identify and eliminate waste from the patient pathways, to enable staff to examine their own workplace, and to increase quality, safety and efficiency in processes (e.g. Fillingham, 2007; Silvester et al, 2004; Radnor and Boaden, 2008).

The barriers specific to the deployment of Lean Six Sigma in health care, in addition to the ones commonly present in other industries, are:

- Measurement: it is often difficult to identify processes, which can be measured in terms of defects (Lanham and Maxson-Cooper, 2003).

- Psychology of the workforce: in the health-care industry it is particularly important to not use jargonistic business language, as this has a high chance of being rejected or accepted with cynicism by medical professionals

The application of Lean Six Sigma in health care is still in its early stages. Therefore early successes in simple projects will pave the way for tackling more complicated initiatives in the future, initiating a positive circle of improvement, bringing clinical change on a broad scale.

Appropriately implemented, Lean Six Sigma can produce benefits in terms of better operational efficiency, cost-effectiveness and higher process quality (Taner et al, 2007), as the case studies presented in this paper illustrate.

The spiralling costs of health care means that unless health-care processes become more efficient, a decreasing proportion of citizens in industrialized societies will be able to afford high-quality health care (de Koning et al, 2006). Continuous process improvement is needed to ensure health-care processes are efficient, cost-effective and of high quality.

The five case study applications we have examined in this paper provide examples of how Lean Six Sigma can help to improve health-care processes. The adoption of similar programs in other hospitals across the health-care sector will help the delivery of high quality health care to an increasing population.

7. Conclusion

Lean Six Sigma is now accepted widely as a business strategy to improve business profitability and achieve service excellence, and its use in service organizations is growing quickly. However, there are a number of barriers to the implementation of Lean Six Sigma in services, such as the innate characteristics of services, as well as the manufacturing origins of Lean Six Sigma that have conditioned service managers to consider them as physical products only. On the other hand, as shown in the case studies, there are a number of advantages for the use of Lean Six Sigma in services (Eisenhower, 1999). Overall, the applications so far have showed the benefits (such as lowering operational costs, improving processes quality, increasing efficiency) to outweigh the costs associated with its implementation.

8. References

Adams, C., Gupta, P. & Wilson, C. (2003) *Six Sigma deployment*. Burlington, MA, Butterworth-Heinemann.

Antony, J. (2005a) Assessing the status of six sigma in the UK service organizations. *Proceedings of the Second National Conference on Six Sigma*, Wroclaw, pp. 1-12.

Antony, J. (2005b) Six Sigma for service processes. *Business Process Management Journal*, 12(2), 234-248.

Antony, J., Antony, F. & Taner, T. (2006), The secret of success. *Public Service Review: Trade and Industry*, 10, 12-14.

Antony, J., Kumar, M. & Cho, B.R. (2007) Six Sigma in services organizations: benefits, challenges and difficulties, common myths, empirical observations success factors. *International Journal of Quality Reliability Management*, 24(3), 294–311.

Antony, J., Kumar, M. & Madu, C.N. (2005) Six Sigma in small and medium sized UK manufacturing enterprises: some empirical observations. *International Journal of Quality & Reliability Management*, 22(8), 860-874.

Antony, J., Kumar, M. & Tiwari, M.K. (2005) An application of Six Sigma methodology to reduce the engine overheating problem in an automotive company. *IMechE – Part B*, 219(B8), 633-646.

Armstrong, M. (2006) *A handbook of human resource management practice*. London, Kogan Page.

Becker, G. S. (1993) *Human capital: a theoretical and empirical analysis, with special reference to education*. Chicago, University of Chicago Press.

de Koning, H., Verver, J. P. S., Van den Heuvel, J., Bisgaard, S. & Does, R. J. M. M. (2006) Lean Six Sigma in health care. *Journal for Healthcare Quality*, 28(2), 4-11.

Eisenhower, E. C. (1999) The implementation challenges of Six Sigma in service business, *International Journal of Applied Quality Management*, 2(1), 1-24

Fillingham, D. (2007) 'Can lean save lives? *Leadership in Health Services*, 20(4), 231-41.

George, M.L. (2003) *Lean Six Sigma for service: how to use Lean speed and Six Sigma quality to improve services and transactions*. New York, McGraw-Hill.

George, M.L. (2002) *Lean Six Sigma: combining Six Sigma quality with Lean speed*. New York, McGraw-Hill.

Gettys, R. (2009) *Using Lean Six Sigma to improve Call Centre operations*. [Online]Available from: http://finance.isixsigma.com/library/content/c070418a.asp [Accessed 22nd January 2009].

Gupta, P. (2005) Six Sigma in HR, *Quality Digest*, QCI International.

Harry, M. and Schroeder, R. (1999) *Six Sigma: The breakthrough management strategy*

International Monetary Fund (IMF), (2011) *World Economic Outlook Database*. [Online] Available   from: http://www.imf.org/external/pubs/ft/weo/2011/01/weodata/index.aspx. [Accessed 7th August 2011]

Jacowski, T. (2008) *Maximizing call centre resource utilization with Six Sigma*. [Online] Available from: http://ezinearticles.com/?Maximizing-Call-Centre-Resource-Utilization-With-Six-Sigma&id=1014905. [Accessed 22nd January 2009].

Jamrog, J. J. & Overholt, M. H. (2005) The future of HR metrics, *Strategic HR Review, 5* (1) 3-3.

Kotler, P. (1997) *Analysis, planning ,implementation and control*, 9th ed. Prentice-Hall.

Lanham, B. & Maxson-Cooper, P. (2003) Is Six Sigma the answer for nursing to reduce medical errors?, *Nursing Economics*, 21(1), 39-41.

Laureani, A. & Antony, J. (2010) Reducing employees' turnover in transactional services: a Lean Six Sigma case study, *International Journal of Productivity and Performance Management*, 59(7), 688-700

Laureani, A., Antony, J. & Douglas, A. (2010) Lean Six Sigma in a call centre: a case study, *International Journal of Productivity and Performance Management*, 59(8), 757-768

Lee, Y., Chen, L. & Chen, S. (2008) Application of Six Sigma methodology in human resources to reduce employee turnover rate: a case company of the TFT-LCD industry in Taiwan. *International Journal of Operations and Quantitative Management*, 14 (2), 117-128.

Milmore, M. et al, (2007) *Strategic human resource management: contemporary issues*. Prentice Hall/Financial Times.

Pande, P., Neuman, R. & Cavanagh, R. (2000) *The Six Sigma way: how GE, Motorola and other top companies are honing their performance.* New York, McGraw-Hill.

Piercy, N. & Rich, N. (2009) Lean transformation in the pure service environment: the case of the call centre. *International Journal of Operations & Production Management,* 29 (1), 54-76.

Radnor, Z. & Boaden, R. (2008) Editorial: does Lean enhance public services?, *Public Money and Management,* 28(1), 3-6.

Regan, W.J. (1963) The Service Revolution, *Journal of Marketing,* 47, 57-62

Rosenberg, A. (2005) Six Sigma: the myth, the mystery, the magic: can Six Sigma really make an impact in your call centre? [Online] Available from http://www.callcentremagazine.com/shared/article/showArticle.jhtml?articleId= 59301130 [Accessed 22nd January 2009].

Silvester, K., Lendon, R., Bevan, H., Steyn, R. & Walley, P. (2004) Reducing waiting times in the NHS: is lack of capacity the problem? *Clinician in Management,* 12(3), 105-11.

Snee, R. D. (2010) Lean Six Sigma: getting better all the time, *International Journal of Lean Six Sigma,* 1(1), 9–29.

Snee, R.D. (2004) Six Sigma: the evolution of 100 years of business improvement methodology. *International Journal of Six Sigma and Competitive Advantage,* 1(1), 4–20.

Snee, R. D. (1999) Why should statisticians pay attention to Six Sigma? *Quality Progress,* 32(9), 100–103.

Sullivan, J. (2003) *HR metrics the world class way,* Kennedy Information.

Swank, C. (2003) The Lean service machine. *Harvard Business Review,* October, 123-129.

Taner, M. T., Sezen, B. & Antony, J. (2007) An overview of Six Sigma applications in the health-care industry. *International Journal of Health Care Quality Assurance,* 20(4), 329-340

Ulrich, D. (1996) *Human resource champions. The next agenda for adding value and delivering results.* Boston, Harvard Business School Press.

Wheeler, D. J. & Lyday, R. W. (1990) *Evaluating the measurement process.* 2nd ed. SPC Press.

Womack, J. P. & Jones, D. T. (1996) *Lean thinking.* New York, Simon & Schuster.

Wyper, B. & Harrison, A. (2000) Deployment of Six Sigma methodology in human resource function: a case study. *Total Quality Management,* 11, (4/5/6), 720-727.

Zeithaml, V.A., Parasuraman, A. & Berry, L.L. (1985), Problems and strategies in services marketing, *Journal of Marketing,* 49 (Spring), 33-46.

2

Utilizing Innovation and Strategic Research and Development to Catalyze Efficient and Effective New Product Development

Yair Holtzman

Director WTP Advisors, Business Advisory Services Practice Leader
USA

1. Introduction

The ability to effectively innovate and develop new products is a vital core competency that any company must possess if it is going to be profitable and experience growth. At the same time, the innovation and effective new product development process is one of the most challenging to nurture and shepherd through to successful completion. It involves a combination of high risk and high return; mission critical importance; and immense scientific, engineering, and financial hurdles. But ultimately, the greatest paradox that surrounds successful new product development is the need for free, unfettered creativity to complement tremendous systematic discipline.

There appear to be three key components to successful strategic innovation and research and development. First, strategic innovation and new product development rest on the effective and efficient use of resources to create clearly differentiated products that customers perceive as more valuable. Second, the process demands a seamless integration and alignment of research and development and new product development objectives and goals throughout the relevant departments of an organization. Third, research and development and new product development efforts should be strategic in nature. This suggests the development of new research and development capabilities, new relationships with customers or suppliers, and more efficient and effective deployment of resources that create lasting competitive advantage for the company.

This chapter will illustrate the importance and interconnection of strategic innovation, research and development, and new product development efforts with clear operations strategy. It will also highlight some of the latest concepts, tools, and strategies in research and development and new product development, displaying why it is vital for firms to establish effective product development processes. Furthermore, it will show how developing this internal capability will permit a company's long term survival and growth in a highly competitive global marketplace.

2. Why is new product development important?

Even though new products require substantial resources, involve high levels of risk, and often result in failure, in many industries, the development and introduction of new products can lead an otherwise faltering company to success. Firms undertake systematic new product development efforts in order to gain competitive advantage, increase market share, reach higher levels of profitability, improve brand equity, and develop new research and development capabilities.

3. Sources of competitive advantage

Firms innovate and develop because new products provide unique opportunities for competitive advantage. Early movers have the advantage of taking a leading role in setting industry standards for emerging product categories. Pharmaceutical companies, for instance, often undertake the simultaneous development of multiple new products because their existing drugs no longer enjoy patent protection. Furthermore, in today's business environment, the products and services produced by a company serve as its face to the public. That is, customers judge a company on its output: great products and services equal a great company. As a result, companies choose to invest significant amounts of time and financial resources in developing new products and services. The innovative product line of Apple, including both the iPhone and iPod, can be seen as instrumental in the survival and emergence of a stronger and more competitive corporation. On the other hand, as the photography industry shifted to a digital focus, Eastman Kodak failed to move forward in creating innovative product offerings. The downfall of the corporation is largely attributed to that decision. As these examples demonstrate, a strong connection exists between how companies go about developing products and services in the marketplace, and the ultimate success or failure of those companies. Accordingly, many seek to understand the evolution of such offerings and learn how to develop innovative products in an efficient and cost effective manner.

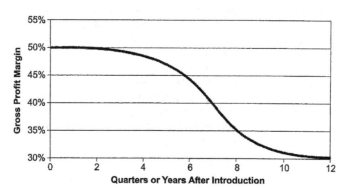

Note: Gross Profit Margin drops over time as commoditization drives prices down

Fig. 1. Cumulative advantage of early entrants (Holtzman, 2010).

4. Market share gain

Over the past several years, two factors have changed the process of gaining market share: an increase in the speed and scale of market and technological changes, and a greater understanding of the interconnectivity of the processes by which products and services are developed and the resulting outcomes of those processes. These factors bring additional focus on the need to be first to market. New products introduced in the marketplace provide additional opportunity for the company to gain first mover advantage. For example, Toyota introduced a hybrid car, the Prius, in advance of its competitors. This has afforded Toyota a dominant position in the fuel-efficient and environmentally friendly automobile market. By developing new products, a company can capture a significant share of the market before competitive products are introduced.

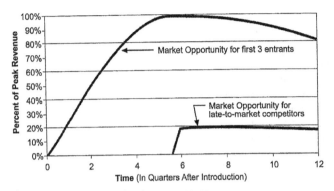

Fig. 2. Gross profit margin over time (Holtzman, 2010).

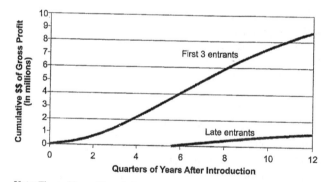

Note: The problem with not being among the first to market with an innovation is clear: late entrants frequently do not earn enough to repay their investment in development, and may be in a significantly weakened condition, unable to fight the next battle

Fig. 3. Early entrant advantage (Holtzman, 2010).

5. Higher profitability

During the early stage, a new product faces less competition than a product in a mature market; therefore, its profitability tends to be higher. This higher profitability, in part, is due to the company capturing a larger share of the market than it would be able to obtain were competition present, as outlined above. It can also partially be attributed to the buying habits of consumers. When a new product is introduced, it will appeal to the innovative crowd. This is particularly true in regard to technological products. From there, if successful, it will eventually reach the main stream, but its popularity, as well as its profit margins, will decrease as new products continue to be introduced and the market becomes saturated with competitive products. This general trend is observed in many industries, and some industries, like the pharmaceutical industry, lock out the competition for several years through patent protection. Patent protection is highly beneficial to pharmaceutical companies because of the extremely high costs and low success rates associated with new product development in that industry.

6. Brand equity

The development of innovative and creative new products is a powerful source of customer loyalty and positive corporate image. Though it remains extremely difficult to quantify the monetary value or goodwill associated with brand image, the imperative nature of strategic actions with this focus is seen in the success that companies like Apple, Google, and BMW have achieved through the results of new product development efforts. While new product development is not the only factor contributing to the image of a company, it is reasonable to assume that substantial brand equity is obtained in this manner. When measured through the use of marketing tools, results show that firms with more successful new product development command higher respect from customers, which leads to enhanced long-term profitability.

7. Resource allocation and development of future research and development

A firm's competitive advantage evolves from its available resources. Resources are physical assets, such as land, equipment, buildings, and cash; intangible resources, such as brand name, market share, product patents, and technological know-how; or capabilities, such as learning proficiencies, product development processes, fast delivery times, and managerial abilities. Analyzing a firm's resources is an important step for a manager to take when formulating and implementing strategy. Valuable resources must support a successful strategy, and when a strategy changes, accompanying adjustments in competitive advantage must be made, which, in turn, will require further resources.

Very often, sustainable sources of competitive advantage evolve into core competencies in a company. Core competencies are capabilities that emerge over time as being central to a firm's overall strategy and upon which the firm's strategy is eventually based. For example, Intel's core competency is the ability to produce the fastest chip in the world. Intel has built its strategy around this core competency for the past decade. More recently, it has begun to stress efficiency in manufacturing because it realizes this will become a more important strategic factor as the industry matures. Shifting its focus will enable Intel to develop new capabilities and competencies that will enable it to continue to compete strategically as a market leader despite changes in the market itself. Multi-business firms build their businesses around a core competency, and this enables them to effectively and efficiently execute a cohesive strategy in all of their businesses.

One of the key strategic decision-making judgments managers face is deciding which resources to develop or acquire. Senior management spends an inordinate amount of time analyzing, selecting, acquiring, and developing the resources necessary in enabling its firm to be competitive. These resources and competencies must be constantly upgraded or altered to enable a firm to maintain its competitive advantage relative to other firms in the market.

8. A mathematical view of the resources allocation problem

Mathematical tools can be used to provide guidance in the deployment of resources in research and development efforts. These tools are only mentioned here in brief, as a detailed discussion is beyond the scope of this text. Linear programming is a widely used mathematical technique designed to help research and development managers plan the optimal allocation of resources.

Requirements of a Linear Programming Problem:

All linear programming problems seek to maximize or minimize some quantity (number of new products brought to market or number of new innovative products or processes). This property is referred to as the **objective function** of a linear programming problem. The major objective of a typical firm is to maximize dollar profits in the long run. In the case of a high technology company, it could be tasked with maximizing the most new innovative products to market in a 6 or 12 month period of time. The presence of restrictions, or **constraints,** limits the degree to which a company can pursue its objective. Therefore, it wants to maximize or minimize a quantity (the objective function) subject to limited resources (the constraints). In order for these calculations to be useful, the company must be deciding among alternative courses of action. For example, if a company can take on three major research and development projects, senior management may use linear programming to decide how to allocate its limited resources (engineers, scientists, lab equipment) among these projects. If there are no alternatives to select among, linear programming is unnecessary. The objective function and constraints in linear programming must be expressed in terms of linear equations or inequalities.

Formulating the Linear Programming Problem:

One of the most common linear programming applications is the product-mix problem. New product development or research and development efforts generally undertake multiple initiatives. Two or more research and development efforts are usually pursued using limited resources. The company would like to determine how many units of research and development products or new products it should produce to maximize overall profit given its limited resources. Note: the reader should not assume that the solution to the linear programming problem is the only variable that needs to be taken into consideration when we discuss research and development efforts. Since new product development and research and development differ from a traditional manufacturing problem, non-quantitative considerations such as strategy and innovation and research and development capability, as well as the "disruptiveness" of the innovation, need to be given careful consideration. The linear programming exercise does provide valuable guidance as to what research and development projects (or activities) potentially have the highest value. The example below is helpful in illustrating some of the basic mathematics in solving a linear programming problem.

The ABC High Technology Company:

The ABC High Technology Company currently has two major research and development projects: (1) the pink R-Bot, a new medical imaging device, and 2) the Chemical BlueBerry, a device capable of detecting noxious chemicals in liquid, gas, or solids with incredible accuracy and needing only tiny samples. The research and development effort for each product is similar in that both require a certain number of hours of electronics engineering and a certain number of hours of chemical engineering. Each pink R-Bot takes 4 hours of electronics engineering and 2 hours of chemical engineering work. Each Chemical Blueberry requires 3 hours of electronics engineering and 1 hour of chemical engineering development efforts. During the current research and development efforts, 240 hours of electronics engineering time are available, and 100 hours of chemical engineering time are available. Each pink R-Bot sold (if successfully produced) yields a profit of $7; each Chemical BlueBerry produced (if successful) may be sold for a profit of $5.

ABC's challenge is to determine the optimal combination of pink R-Bot's and Chemical BlueBerrys to maximize profit. Assume for this example that both new products share the same viability and market success. This new product and research and development mix of resources situation can be formulated as a linear programming problem.

The table below summarizes the information to formulate and solve this problem (see Table Ex. 1). Furthermore, let's introduce some simple notation for use in the objective function and constraints.

Let:

- X_1 = the number of pink R-Bots to be developed
- X_2 = the number of Chemical BlueBerries to be developed

Engineering Department	Pink R-Bot	Chemical BlueBerry	Available Hours/Week
Electronics	4	3	240
Chemical	2	1	100
Profit per Unit	$7	$5	

Table 1. Hours required to Develop One Unit of New Research & Development Effort.

Now we can create the LP objective function in terms of X_1 and X_2:

$$\text{Maximize Profit} = \$7X_1 + \$5X_2$$

Our next step is to develop mathematical relationships to describe the two constraints in this problem. One general relationship is that the amount of a resource used is to be less than or equal to ($\leq$) the amount of resource available.

First constraint:

Electrical engineering time required is $\leq$ Electrical engineering time available.

$$4X_1 + 3X_2 \leq 240 \text{ (Electrical Engineering time available)}$$

Second constraint:

Chemical engineering time required is $\leq$ Chemical engineering time available.

$$2X_1 + 1X_2 \leq 100 \text{ (Chemical Engineering time available)}$$

Both of these constraints represent possible research and development resource capacity restrictions and certainly will affect the total profit outcome. For example, ABC Technology

Company cannot develop 100 pink R-Bots because if $X_1 = 100$, both constraints would be violated. ABC could also not allocate its resources such that $X_1 = 50$ and $X_2 = 10$. This constraint illustrates that interactions exist between variables. The more units of one new product that are worked on, the fewer resources ABC Technology has to allocate to the other research and development efforts.

The easiest method to solve a small LP problem such as that of ABC Technology Company is the graphical solution approach. The graphical procedure can be used only when there are two decision variables such as in our example. When there are more than two variables, it is not possible to plot the solution on a two dimensional graph; we must then turn to more complex approaches (see other books on Management Science or Operations Research).

Graphical Representation of Constraints:

To determine the optimal solution to a linear programming problem, we must first identify a set, or region, of feasible solutions. This is solved by plotting the problem's constraints on a graph. The variable X_1 (pink R-Bot) is usually plotted as the horizontal X- axis of the graph, and the variable X_2 (Chemical BlueBerry) is plotted as the vertical, Y-axis. The complete problem we want to solve then is:

Maximize Profit (New Product Development) = $\$7X_1 + \$5X_2$ subject to the constraints:

$$4X_1 + 3X_2 \leq 240 \text{ (Electrical Engineering time available)}$$

$$2X_1 + 1X_2 \leq 100 \text{ (Chemical Engineering time available)}$$

Note that $X_1 > 0$ and $X_2 > 0$. These last two constraints are also called the non-negativity constraints. The first step in graphing the constraints is converting the constraint inequalities into equalities or equations.

Example--Figure 1: $4X_1 + 3X_2 = 240$ (Electrical Engineering time available)

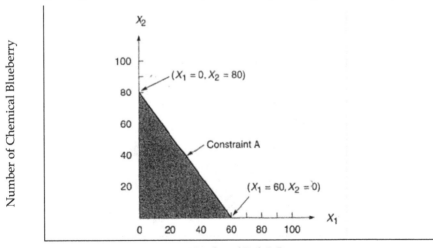

Fig. 1. Constraint 1.

Example-- Figure 2: $2X_1 + 1X_2 = 100$ (Chemical Engineering time available)

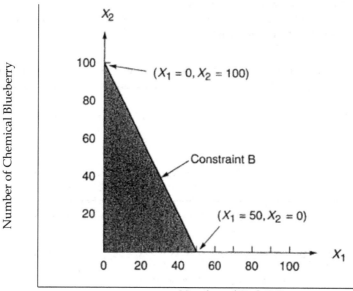

Fig. 2. Constraint 2.

Example-- Figure 3 below shows both constraints together. The shaded region is the part that satisfies both restrictions. The shaded region in Figure 3 is called the area of feasible

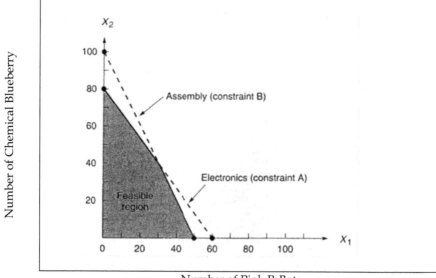

Fig. 3.

solutions, or simply the feasible region. This region must satisfy all conditions specified by the program's constraints and is therefore the region where all of the constraints overlap. Any point in the region would be a feasible solution to the ABC Technology company question of where best to allocate the scare hours of the chemical and electronics engineers. Furthermore, any point outside the shaded area would represent an infeasible solution.

Now that the feasible region has been graphed, we can proceed to find the optimal solution to the problem. The optimal solution is the point lying in the feasible region that produces the highest profit, or produces the greatest number of new products, or the greatest number of "disruptive innovations". Once the feasible region has been established, several approaches can be taken in solving for the optimal solution. One method used is called the iso-profit line method. See Example — Figures 4 & 5 below.

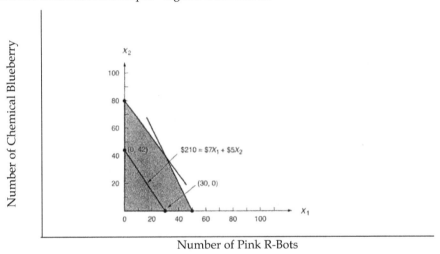

Fig. 4. A Profit Line of $210 Plotted for ABC Technology Company.

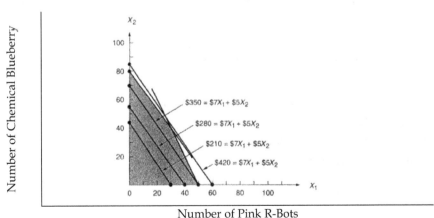

Fig. 5. Four Iso-Profit Lines Plotted for ABC Technology Company.

The last method that should be mentioned is the four corners of the feasible region method. This is shown in Example—Figures 6 & 7 below. In this method it is necessary to find only the values of the variables at each corner; the maximum benefit or optimal solution will lie at one (or more) of them. Once again we can see that (in Example—Figure 7) that the feasible region for ABC Technology Company problem is a four-sided polygon with four corner or extreme points. The points are labeled 1, 2, 3 and 4 on the graph. To find (X_1, X_2) values producing the maximum profit, we find out what the coordinates of each corner point are, then determine and compare their benefit levels.

Point 1: $(X_1 = 0, X_2 = 0)$ Benefit = 0

Point 2: $(X_1 = 0, X_2 = 80)$ Benefit = $7(0) + $5(80) = $400

Point 4: $(X_1 = 50, X_2 = 0)$ Benefit = $7(50) + $5(0) = $350

Corner point #3 needs to be solved algebraically. We apply the method of solving simultaneous equations to the two constraint equations:

$4X_1 + 3X_2 \leq 240$ (Electrical Engineering time available)

$2X_1 + 1X_2 \leq 100$ (Chemical Engineering time available)

To solve these simultaneous equations, we multiply the second equation by -2 to obtain:

$-2(2X_1 + 1X_2 = 100)$ (Chemical Engineering time available) $= -4X_1 - 2X_2 = -200$

And solve the two equations to obtain $X_2 = 40$ and therefore X_1 should equal 30. Therefore, point 3 produces the highest profit of any corner point, and point 3 has coordinates $(X_1 = 30, X_2 = 40)$ Benefit = $7(30) + $5(40) = $410. Because point 3 produces the highest profit of any corner point, the optimal mix of resource allocation suggests making resources available for 30 units of pink R-Bots and 40 units of resources available for Chemical Blueberries, which solves ABC Technology Company's scarce resource allocation problem.

Another approach to solving linear programming problems utilizes the corner-point method. This technique involves looking at the profit at every corner point of the feasible region. The mathematical theory behind linear programming states that an optimal solution to any problem (that is the values of X_1, X_2 that yield the maximum variable of interest, which is generally profit, but could just as easily be new product introductions) will lie at a corner point, or extreme point of the feasible region. Therefore, it is necessary to find only the values of the variables at each of the corners. The variable of interest will be maximized at one of the corner points (see mathematical discussion and figure above).

There are two problems that arise in the deployment of scarce resources. The first is the 1) Activity Analysis Problem and 2) The Optimal Assignment problem.

The Activity Analysis Problem. There are n activities, A_1, A_2. .. . ,An ,that a company may employ, using the available supply of m resources, R_1, R_2. . . . , Rm (labor hours, steel, etc.). Let bi be the available supply of resource Ri. Let aij be the amount of resource Ri used in operating activity Aj at unit intensity. Let cj be the net value to the company of operating activity Aj at unit intensity. Choose the intensities with which the various activities are to be operated to maximize the value of the output to the company subject to the given resources.

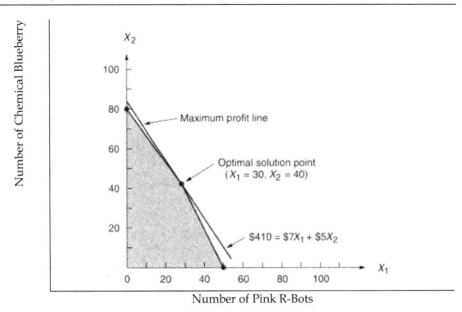

Fig. 6. Optimal Solution for ABC Technology Company.

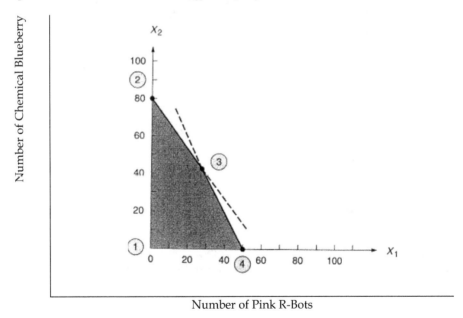

Number of Pink R-Bots

Fig. 7.

Let xj be the intensity at which Aj is to be operated. The value of such an activity allocation
is

$$\sum_{j=1}^{n} c_j x_j. \tag{8}$$

The amount of resource Ri used in this activity allocation must be no greater than the supply, bi; that is,

$$\sum_{j=1} a_{ij} x_j \le b_i \qquad \text{for } i = 1, \dots, m. \tag{9}$$

It is assumed that we cannot operate an activity at negative intensity; that is,

$$x_1 \ge 0, x_2 \ge 0, \dots, x_n \ge 0. \tag{10}$$

Companies want to: maximize (8) subject to (9) and (10). This is exactly the standard

maximum problem that a company with scarce research and development resources needs to address and optimize in its desire to advance its innovation and research and development capabilities. Another critical issue that needs to be addressed with scarce resources, is that of optimally assigning these resources.

The Optimal Assignment Problem. There are I persons available for J jobs. The value of person i working 1 day at job j is aij , for i = 1, . . . , I, and j = 1, . . . , J . The problem is to choose an assignment of persons to jobs to maximize the total value.

An assignment is a choice of numbers, xij , for i = 1, . . . , I, and j = 1, . . . , J, where xij represents the proportion of person i 's time that is to be spent on job j. Thus,

$$\sum_{j=1}^{J} x_{ij} \le 1 \qquad \text{for } i = 1, \dots, I \tag{11}$$

$$\sum_{i=1}^{I} x_{ij} \le 1 \qquad \text{for } j = 1, \dots, J \tag{12}$$

and

$$x_{ij} \ge 0 \qquad \text{for } i = 1, \dots, I \text{ and } j = 1, \dots, J. \tag{13}$$

Equation (11) reflects the fact that a person cannot spend more than 100% of his time working, (12) means that only one person is allowed on a job at a time, and (13) says thatno one can work a negative amount of time on any job. Subject to (11), (12) and (13), we wish to maximize the total value,

$$\sum_{i=1}^{I} \sum_{j=1}^{J} a_{ij} x_{ij}. \tag{14}$$

This is a standard maximum problem with m = I + J and n = IJ .

9. Terminology

The function to be maximized or minimized is called the objective function. A vector, x for the standard maximum problem or y for the standard minimum problem, is said to be feasible if it satisfies the corresponding constraints.

The set of feasible vectors is called the constraint set.

A linear programming problem is said to be feasible if the constraint set is not empty; otherwise it is said to be infeasible.

A feasible maximum (resp. minimum) problem is said to be unbounded if the objective function can assume arbitrarily large positive (resp. negative) values at feasible vectors; otherwise, it is said to be bounded. Thus there are three possibilities for a linear programming problem. It may be bounded feasible, it may be unbounded feasible, and it may be infeasible.

The value of a bounded feasible maximum (resp, minimum) problem is the maximum (resp. minimum) value of the objective function as the variables range over the constraint set. A feasible vector at which the objective function achieves the value is called optimal.

All Linear Programming Problems Can be Converted to Standard Form.

A linear programming problem was defined as maximizing or minimizing a linear function subject to linear constraints. All such problems can be converted into the form of a standard maximum problem by the following techniques.

A minimum problem can be changed to a maximum problem by multiplying the objective function by -1. Similarly, constraints of the form $\sum_{j=1}^{n} a_{ij}x_j \geq b_i$ can be changed into the form $\sum_{j=1}^{n} (-a_{ij})x_j \leq -b_i$. Two other problems arise.

(1) *Some constraints may be equalities.* An equality constraint $\sum_{j=1}^{n} a_{ij}x_j = b_i$ may be removed, by solving this constraint for some x_j for which $a_{ij} \neq 0$ and substituting this solution into the other constraints and into the objective function wherever x_j appears. This removes one constraint and one variable from the problem.

(2) *Some variable may not be restricted to be nonnegative.* An unrestricted variable, x_j, may be replaced by the difference of two nonnegative variables, $x_j = u_j - v_j$, where $u_j \geq 0$ and $v_j \geq 0$. This adds one variable and two nonnegativity constraints to the problem.

Any theory derived for problems in standard form is therefore applicable to general problems. However, from a computational point of view, the enlargement of the number of variables and constraints in (2) is undesirable.

New Product Development Defined:

The new product development process can be defined as a good, service, or a good/service package, which was previously unavailable to customers, becoming available to the marketplace. From the perspective of a company, a new product also can be offering a good or service the company did not previously offer.

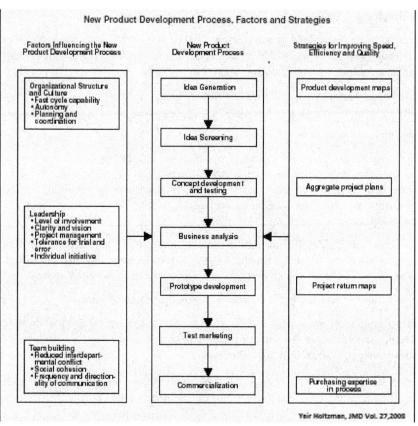

Fig. 8. New product development process, factors and strategies.

Effective research and development requires close and highly integrated links with many different parts of an organization that produces the product or service. Seamless integration and cooperation of various departments are essential to strategic new product development operations. Many new product ideas are based on existing products and are developed from within the production or service operations. It is not uncommon for a company's research and development department to develop a new product, and for various reasons, the manufacturing department is unable to produce the product in an efficient and effective manner. An effective product development strategy links product development decisions with cash flow, market dynamics, product life cycle, and the organization's overall capabilities. As a result, the screening process should extend to the operations function. Identifying products that are likely to capture market share, be innovative, be cost effective, and be profitable, but are in fact very difficult to produce, can lead to disappointment and failure rather than success. Motorola, for example, went through over 3100 working models before it arrived at its first working pocket cell phone. Optimal product development and innovation depends not only on support from other parts of the firm but also critically upon the successful integration of operations management decisions, from product design to maintenance.

Transparency and alignment of objectives and goals across all departments working on a research and development project is critical to the project's success. In general, it will enhance creativity and ownership by all, and reduce redundancies and inefficiencies. Figure 8 summarizes several of the new product development challenges and some of the required ownership "buy-in" required to tackle these.

10. Research and development defined

Research and development is sometimes referred to as "engineering" in the Life Sciences field since the design and development of medical devices is heavily dependent on activities related to one or more of the engineering disciplines. Research and development typically includes all engineering and testing activities beyond early-stage prototyping and final concept selection to the point when a product is ready to be released into production. A company's ability to transform an initial "proof of concept" prototype into a final product is central to its viability. The research and development process is critical for the success of companies that manufacture a technology product for several other tactical and strategic reasons as well. From a practical short term perspective, research and development:

1. Plays an instrumental role in defining how the original need is ultimately addressed.
2. Provides the engineering framework for developing a company's technology with the least amount of risk.
3. Facilitates the management of a primary driver of cost early in the company's life in terms of how personnel and other resources are utilized and managed.
4. Provides the foundation (processes, technology capabilities, and culture) that enables a company to continue innovating and developing future product iterations.
5. Often leads to critical insights related to the firm's intellectual property position.

From a longer term perspective, a strategic approach to research and development can also help a company:

1. Continually increase product differentiation, thereby solidifying market position and mitigating risk of competition.
2. Drive growth through new product innovation.
3. Develop new R&D capabilities.
4. Create a product development pipeline, which can make the company more attractive to investors and/or prospective acquirers.
5. Discover new markets and new cross-selling product capability.

There are various ways through which a company can decide to perform its research and development efforts. A study of the field of strategic research and development and new product development suggests that the most successful organizations use a research and development strategy that ties external opportunities to internal strengths and is linked with well-defined objectives. Well-formulated and executed research and development policies match market opportunities with internal capabilities and provide an initial screen for all ideas generated. Research and development policies can enhance strategy implementation efforts by either:

1. Performing research and development within the firm or contracting research and development to outside firms.

2.   Using university researchers or private sector researchers.
3.   Emphasizing product or process improvements.
4.   Stressing basic or applied research.
5.   Becoming leaders or followers in research and development.

In implementing different types of generic business strategies, effective and efficient interactions must exist between R&D departments and other functional departments. Lack of synchronization between marketing, finance and accounting, manufacturing, research and development, and operations departments can and should be minimized with clear policies and objectives.

Many firms struggle with the decision whether to acquire research and development expertise from external firms or to develop expertise internally. Based on my experience servicing companies in developing and harnessing their research and development capabilities, many companies have implemented the guidelines below relatively successfully as they make this decision:

1. In an industry where the rate of technical progress is slow, the rate of market growth is moderate, and significant barriers limit possible new entrants, in-house research and development is the preferred solution. By working internally, successful research and development will result in a temporary product or process monopoly the company can exploit.

2. If technology is changing rapidly and the market is growing slowly, a major effort in research and development can be very risky because it ultimately could lead to development of an obsolete technology or one for which there is no market.

3. If technology is changing slowly but the market is growing rapidly, there generally is not enough time for in-house development. The prescribed approach is to obtain research and development expertise on an exclusive or nonexclusive basis from an outside firm.

4. If both technical progress and market growth are occurring rapidly, research and development expertise should be obtained through the acquisition of a well-established firm in the industry.

A firm can take at least three different approaches to implementing research and development strategies. It can become:

1.   The first to market new technological products.
2.   An innovative imitator of successful products.
3.   A low-cost producer of similar but less expensive products.

Of the various strategies, being the first firm to market new technological produces seems most glamorous and exciting. Nevertheless, this can be a risky approach, particularly if the market does not appreciate the new differentiated product. By serving as an innovative imitator of successful products, a firm is able to minimize risks and start-up costs. This approach entails allowing a pioneer firm to develop the first version of a new product and demonstrate that a market exists. Laggard firms then develop a similar product. This strategy requires excellent research and development personnel and a strong marketing department.

A low-cost producer mass-produces products similar to, but less expensive than, products recently introduced. Recent trends indicate that research and development management have introduced an additional approach as some firms have begun to lift the proverbial veil of secrecy. In some instances, major competitors are joining forces to develop new products. Collaboration is on the rise due to new competitive pressures, rising research costs, increasing regulatory issues, and accelerated product development schedules.

11. Strategic innovation

Many organizations rely on random acts of creativity and innovation, which is not the way to drive successful innovation in the long run. In my experience, companies are best served by nurturing and facilitating a **strategic innovation approach.** Strategic innovation takes place on several levels and includes both traditional and non-traditional approaches to business strategy. First, it should consist of industry knowledge and foresight. This effort seeks breakthrough disruptive innovation while continuing to build core competencies. Secondly, a company should seek to understand its customer, provide insight, pinpoint unarticulated customer needs and delights, and deliver on these unmet needs. Lastly, strategic alignment should occur both internally, within the company, and externally, with customers and suppliers. This involves ensuring that all departments are aligned in terms of objectives. Successful strategic innovation involves exploring long term possibilities and practical implementation activities that lead to short term, measurable business benefit.

Strategic innovation is a holistic, systematic approach focused on generating disruptive and discontinuous innovations. Innovation becomes strategic when it is an intentional, repeatable process that creates a significant difference in the value delivered to consumers or the company. A strategic innovation approach initiative generates a portfolio of breakthrough business growth opportunities using a disciplined yet creative process. Based on my findings, seven key components lead to successful strategic innovation. First, managing the innovation process involves selectively combining traditional and non-traditional approaches to business strategy. Second, industry knowledge and foresight means the key decision makers and technology leaders within a company understand the complex forces driving change, including emerging and converging trends, new technologies, competitive dynamics, and competition. Third, strategic alignment means building support for the initiatives from the senior leadership team, within the company, and with any related stakeholders. Fourth, customer insight, including understanding your customer, what they value, and what delights them, is critical to the success of any new product development initiative. Fifth, technologies and competencies are the set of internal capabilities, organizational competencies, and assets that can be leveraged to deliver value to customers. These assets include technologies, intellectual property, brand equity, and strategic relationships. Sixth, the buy-in of the leadership of the entire organization is essential for implementation of strategic innovative ideas to take root. Seventh, effective and efficient implementation of a product from inspiration to completion requires disciplined follow through at all phases of the project. It is challenging to develop creative, visionary thinking; it is far more difficult to successfully implement that thinking in a way that creates meaningful business impact.

The managed innovation process extends and covers all of the activities from initial brainstorming sessions through implementation. When we discuss strategic innovation, the term "implementation" includes a wide range of activities. These include transition to specific projects, reallocation of resources, properly incentivizing key employees, and creating an environment that encourages creativity and innovation and does not penalize employees for failure. Implementation demands span well-beyond the scientific and technology departments to encompass technical product testing, value proposition development, a clear marketing campaign, consumer-based rapid prototyping and testing, brand development, business case construction, effective marketing channels, a supply chain, and broad organizational buy-in. In practice, this can take many forms. My experience has included many companies that utilize approaches involving facilitated workshops sessions. These sessions force cross-functional teams to look beyond the expected next step. Implementation can mean successfully bringing a new product to market or increasing market share, which is achieved by utilizing a new development strategy or creating a more robust technology; regardless, the process needs to be carried through to completion. In the process, companies are able to seek out disruptive creative ideas that have the potential to cross pollinate and lead to new levels of creativity.

In order for the managed innovation process to have a chance at success, both the senior leadership team must buy-in completely and all affected parties should contribute. This complete representation will consist of a broad cross-section of the company and any other stakeholder in the supply chain that shares the same vision or desire for the project to succeed. I have seen this work successfully in the biotechnology industry and fail miserably in the automotive industry. The key difference was that the biotechnology company had clearly aligned all critical stakeholders, internal and external, and in the automotive example, the tier one and tier two suppliers were not brought into the fold on these new innovative projects until they were actually provided with a finished product. These tier one and tier two suppliers were then asked to incorporate the new component product on their production lines and send the finished assembled product back to Detroit. There were problems in design that did not allow the product to be ready for mass production, and the product underwent several rounds of iterations.

Active cross-functional participation in the new product development process builds strategic alignment and buy-in among key stakeholders both within an organization and externally. This alignment strengthens the organizational buy-in, creates ownership and excitement, accelerates any decisions that need to be made along the way, and facilitates successful implementation. Strategic alignment is absolutely critical for operational success as it enables cross-functional decisions and agreement on difficult issues surrounding implementation activities, such as resource allocation, competencies, and ownership of parts of the project. Furthermore, for the project to be successful, it is essential that stakeholders are continuously engaged in a meaningful manner from the inception of the idea throughout the implementation process.

One of the most important components of successful strategic innovation is for company leadership to understand the drivers, trends, enablers, and threats to the industry in which the innovation is being undertaken. Visionary organizations typically establish a process for monitoring the complex interplay of key trends that could potentially impact their business. Best-in-class innovators know how to evaluate the various forces in the global marketplace, and as a result, expose potential opportunities in the "white space" and develop robust and effective strategic innovation.

12. Product differentiation: what does it mean and why is it important?

Product differentiation exists when customers genuinely perceive a certain firm's product to be more valuable than other firms' products. For example, customers who purchase BMWs and believe that BMW builds the "ultimate driving machine" are purchasing an automobile that they perceive to be superior to all others on the market. Although differentiation can have several different bases, in the end it is always a matter of customer perception. The attributes of product differentiation include characteristics of the product or service provided the relationship between a firm and its customers, and connections and interrelationships between firms. The last point might not be as clear as the first two; it relates to the firm's product mix, distribution system, supply chain, and level of customer service and support. A company's first line managers maximize product differentiation through creativity, coordination, and transparency.

Why is product differentiation valuable? One reason is that it enables a firm to set its prices higher than it would otherwise be able to charge its customers.

Each of the bases of product differentiation identified can be used to exploit environmental opportunities, capture market share, or even establish the company as a dominant leader in that space. The rarity and uniqueness of bases of product differentiation differ widely. Highly imitable bases of product differentiation include product features. Somewhat imitable features include product mix, interconnections with other firms, and product customization. Extreme difficulty lies in imitating bases of product differentiation that include disruptive technologies or new-to-world products or capabilities. Organizations that have strategically integrated product differentiation throughout the company do indeed develop capabilities that are very robust and difficult to imitate.

Implementation of a product differentiation strategy involves connecting technical efforts to the business as a whole, which is challenging for research and development management of large companies. In general, integrated and synchronized management of the organizational structure, management controls, and well defined and flexible incentive policies can help foster successful strategic new product development and research and development. It is very common for companies implementing product differentiation strategies to use cross-divisional and cross-functional teams in addition to teams focused exclusively on a particular product development effort. Based on my findings, managerial controls that provide free managerial decision making within broad decision-making guidelines can be helpful in providing robust implementation of product differentiation strategies. Furthermore, employees that are incentivized to take some risks through compensation policies that encourage creativity and innovation are more successful in bringing product differentiation strategies and products to market.

For many years it was thought by leaders in the innovation and new product development world that a company had to choose between unique, innovative, top notch products and products that are relatively inexpensive. I spent several years consulting to the automotive industry where, in general, it was thought that plants could not simultaneously build low-cost and high quality automobiles. Between 1994 and 2004, I had the opportunity to visit 54 automotive plants throughout the world that assembled either completed automobiles or large components that go directly into the completed automobile. What I observed at the time the research was done was that there were four plants that were able to have both low

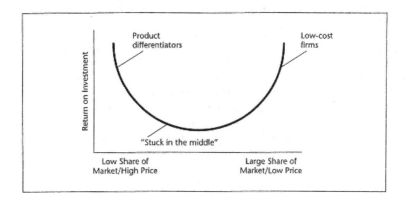

Fig. 9. Simultaneous implementation of cost leadership and product differentiation competitive strategies: being "Stuck in the Middle".

Cost leadership	Product differentiation
Organizational structure	**Organizational structure**
1. Few layers in the reporting structure	1. Cross-divisional/cross-functional product development teams
2. Simple reporting relationships	2. Willingness to explore new structures to exploit new opportunities
3. Small corporate staff	3. Isolated pockets of intense creative efforts
4. Focus on narrow range of business functions	
Management control systems	**Management control systems**
1. Tight cost-control systems	1. Broad decision-making guidelines
2. Quantitative cost goals	2. Managerial freedom within guidelines
3. Close supervision of labor, raw material, inventory, and other costs	3. Policy of experimentation
4. A cost leadership philosophy	
Compensation policies	**Compensation policies**
1. Reward for cost reduction	1. Rewards for risk-taking, not punishment for failures
2. Incentives for all employees to be involved in cost reduction	2. Rewards for creative flair
	3. Multidimensional performance measurement

Fig. 10. The organizational requirements for implementing cost leadership and produce differentiation strategies.

costs and very high quality. I am confident that if I visited all 54 plants today, a larger percentage would be able to deliver both low cost and very high quality. The reason I believe that this tradeoff is becoming less prevalent than it was twenty years ago is primarily for the following three reasons. First, plants are employing best manufacturing technology and practices (Six Sigma, lean manufacturing, computerized robots, laser guided paint machines, etc.). This is coupled with highly participative, group oriented team ownership, including participative management, team production, and total quality management. Second, employees have a clear sense of ownership and take pride in their work product. Third, the employees demonstrate a tremendous amount of loyalty and commitment toward the plant they work for, a sense of loyalty and pride that translates to fewer errors and overall best quality and better work environment.

These four plants had clearly demonstrated that the traditional tradeoff assumed in the industry should not necessarily be taken for granted. Firms can simultaneously implement cost leadership and product differentiation strategies if they learn how to manage the contradictions inherent in these two strategies. Successful management of these two seemingly contradictory strategies depends upon working through socially complex relations among employees, between employees and the technology they use, and between employees and the firm they work for. Successfully managing these challenges can result in both of these strategies complementing each other and fueling further cost reductions and increased differentiation. Connecting scientific and engineering efforts to business objectives is a significant challenge for research and development and high technology manufacturing management. Technical employees who are brought into the fold and the "know" are able to make good decisions, and they may be further motivated to contribute all of their knowledge and creativity and support organizational objectives far better than they otherwise would.

13. The strategic importance of product and service development

As illustrated in Figure 11, product and service development is viewed as increasingly important from a strategic point of view. From a market perspective, international competition has become increasingly intense. In many markets, there are a number of competitors bunched together in terms of their product and service performance. As a result, even small advantages in product and service specifications can have a differentiating impact on competitiveness and ultimately, product survival. This has made customers much more sophisticated in exercising their choice and often more demanding in terms of wanting products and services that fit their specific needs. Furthermore, markets are becoming more fragmented. Unless companies choose to follow relatively narrow niche markets, they are faced with developing products and services capable of being adapted in different ways to different markets. To further exacerbate the technical challenges, product and service life cycles have become shorter. Therefore, introducing new products and services in an efficient and effective manner allows companies to have an advantage over the competition. Because competitors respond by doing the same, the situation escalates.

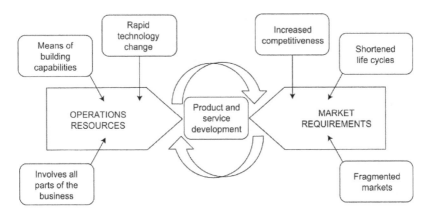

Fig. 11. The increasing strategic importance of product and service development.

An additional set of pressures affect the operations resources that develop and deliver new products and services. Perhaps most importantly, rapid technology changes have affected most industries. Primarily because of the scale and pace of such technological developments, it has become increasingly obvious that effective product and service development places responsibility on every part of the business. Marketing, purchasing, accounting, and operations are all an integral part of the organization's ability to develop products and services effectively and efficiently. Every part of the business is now faced with the question of how it can deploy its particular competencies and skills toward developing innovative, value-adding products and services in an efficient and effective manner.

Stages of Development

The way in which organizations develop products and services is as varied as the products and services themselves. Furthermore, what companies specify as a formal product or service development methodology, as compared to what happens in reality, are usually very different things. Nevertheless, in my experience, the ideas below seem to have found wide acceptance amongst product development companies.

Figure 12 below outlines the development process as it moves through a series of stages. As development progresses through the various stages, some steps might be unnecessary on certain projects, while other steps are repeated multiple times. At the beginning of this process, there are stages concerned with collecting ideas and generating product and service concepts, and toward the end of the process, there are stages concerned with specifying the detail of product or service specifications.

As the development process moves through these stages, the number of alternative design options is reduced until one final design remains. The process often includes decision points that screen out options viewed as unsatisfactory.

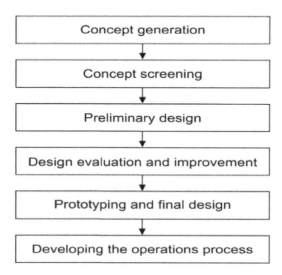

Fig. 12. A typical "stage model" of the product and service development process.

The possible design options are then reduced to a very small set of possible research and development outcomes, and engineers move from a state of uncertainty to a state of increasing certainty. One consequence of this is that the ability to change the design set gets increasingly difficult and limited. Making changes at the end of the development process can be considerably more expensive than making them at the beginning of the development process. Therefore, adding value early on in the research and development process is critical to success.

14. Creating efficiency in research and development and adding value early in the process

In their ongoing desire to become more efficient, many research and development organizations largely focus on reducing administrative inefficiencies, such as burdensome approval processes, lack of information, and meetings that are unproductive or unnecessary. Yet, they tend to overlook the bigger issue of engineering inefficiencies, which are caused by problems such as shifting design requirements, poor integration of design components, and post-production design changes. Although engineering and research and development inefficiencies consume substantial time and resources -- and have a much bigger impact on the bottom line -- they are often ignored because they are more complex and challenging to address. In some cases, they are not even recognized as a problem.

Should companies focus more attention on reducing engineering and research and development inefficiencies and adding more value early in the design process? Or should they accept the status quo and focus their efforts elsewhere?

Here is the debate that many business leaders and research and development executives are wrestling with. Should they focus more attention on engineering waste and accelerate engineering value? The best way to improve engineering efficiency is to avoid unnecessary design changes, minimize rework, and improve coordination between design and manufacturing from the get-go. Reduced engineering waste will lower design costs, accelerate development, and improve overall competitiveness.

Many research and development organizations do not view manageable inefficiency as a waste, but instead as a value-added activity necessary to get the design right. Nevertheless, based on my experience, excessive engineering effort is a clear indicator of inefficiency and can be reduced sharply without an adverse effect on design outcomes.

For example, poorly defined requirements early in the design cycle can cause excessive low value adding effort in later phases, which ultimately increases costs and slows work down. Other common causes of manageable excessive engineering effort include:

Lack of integration. R&D organizations are becoming more and more sophisticated in their use of computer-aided tools to design and model product parts. However, disconnects still can occur when individual design efforts are not tied into a requirements management process that ensures the separate components will ultimately work together. Not having the right tools, processes, people, or data to achieve the necessary integration can contribute to excessive non-value adding activities.

Poor synchronization across design groups. Different design teams tend to work at different speeds. Unless work is scheduled and prioritized, some groups inevitably fall behind or find themselves waiting on others. This is particularly challenging when the time comes to test that various components work together – especially if some components are mechanical while others involve software or are electrical.

Design by committee. Building consensus around decisions can be a valuable exercise. At the same time, excessive deliberation and lack of clearly delineated decision-making roles is counter-productive and can make it difficult or impossible to meet product development deadlines.

Lack of cross-functional integration. Research and development needs to bring other functions into the development process as early as possible. For example, failure to get the manufacturing organization involved can lead to inadequate tools and shop floor processes. Similarly, failure to involve the after-market service organization can create costly support problems for the company and its customers. The best design in the world is useless if it cannot be built and properly supported.

One reason engineering inefficiency is hard to fix is because it is hard to see. The right diagnostic tools can help decision-makers visually analyze engineering data to better determine how much engineering effort is unavoidable and how much can be eliminated. Tools and techniques from lean manufacturing and Six Sigma can then be applied to help address the root causes of manageable churn and add more engineering value in the early stages of the design cycle.

Manufacturing businesses around the world have been operating in a severe cost-cutting environment for more than a year, and the pressure to keep costs down is unlikely to abate any time soon. In fact, for many global manufacturers, cost reduction and limited research and development funding have become a basic business requirement.

Of course, most manufacturing companies do not have the luxury of cost-cutting their way to prosperity and growth. At some point, they will need to invest in new products, markets, and growth opportunities. Achieving sustainable growth and innovation in the face of limited research and development funding requires improved capabilities. Sticking with the status quo is simply not an option.

Attacking engineering inefficiency to significantly improve high value add research and development efficiency, while, at the same time, improving research and development performance, must be considered. For example, a global specialty chemical manufacturer recently identified 50,000 engineering labor hours that could be better spent developing new products. Similarly, a global automotive manufacturer identified millions of dollars in annual engineering labor that could be used more efficiently and eliminated over a dozen of steps in quality control and routine testing.

Identifying and reducing engineering churn can enable a company to redirect wasted resources and effort to activities that create value for the business and improve its overall competitiveness in the marketplace.

Automotive, process, and industrial products manufacturers share a common characteristic: they all manufacture highly engineered products. For these businesses, the benefit opportunities of attacking waste in engineering are the same as for attacking waste in manufacturing operations — reduced cycle time, increased throughput, higher product quality, and lower costs.

Unfortunately, the drivers and impact of non-value activities and inefficiencies in engineering tend to be less apparent than in manufacturing operations, where physical scrap, inefficient flows, and poor production quality often are readily visible. The good news is that analytical tools are available to help companies in their efforts to identify and address the damaging effects of engineering waste.

The payoff for auto, process, and industrial manufacturing companies can be relatively high since engineering costs comprise such a large percentage of their overall cost structure. For example, a global automaker recently identified hundreds of man-months of wasted development time per full vehicle program, which represented millions in excess research and development costs. That is a level of waste and inefficiency that today's manufacturers simply cannot afford if they are to survive in the highly competitive global manufacturing environment.

15. Marketing's role in strategic research & development optimization

Marketing plays a critical role in the development, success, and optimization of strategic research and development in new product and process development.

All businesses face different sets of challenges: challenges in internal operations, in the industry, in the economy, in the marketplace, and in growth stages. Starting a new business can be difficult, time-consuming, and risky. Marketing research and strategy development at this stage are as important as raising finance for the business. A growing business, in addition to managing the growth process, must deploy and leverage new technologies while developing strategies to increasing its share of the market. It must juggle the developing of brands, the management of cash flow, and the development of an effective distribution and supply chains. Established businesses must develop new income streams to sustain profitability in a rapidly changing competitive market, consistently define new markets and develop synergistic partnerships.

Historically, in the pharmaceutical industry, customer-led research and development has not been practiced so rigorously. Pharmaceutical companies set research and development priorities based on the opportunity for scientific discovery combined with long-term revenue forecasts- notably, not profit forecasts- that promise attractive commercial gains. They seek the customer's input- from physicians, payers, and patients- usually only after a product reaches the late-stage pipeline; even then, the feedback influences only launch strategies and market positioning. Moreover, such customer input is heavily focused on physicians, such as the factors that influence their prescriptions. The payer's perspective is largely restricted to reimbursement negotiations. It is rarely an input for setting research and development priorities, and almost never in the early stages of the pipeline, in the labs and clinics. In effect, pharmaceutical companies seldom undertake a rigorous assessment of what payers will be willing to pay for compared to alternative treatments before deciding what to research.

In the future, pharmaceutical companies will need to listen early in their research and development efforts to the voice of customers, especially payers. While a pharmaceutical company cannot design products tailored to customer specs, as SAP does, it can guide its research and development closer to customer needs. That shift is imperative. As payers consolidate, they are becoming more powerful and cost conscious, demanding hard evidence that their reimbursement dollars are well spent. By identifying which health outcomes payers are more willing to reimburse, pharmaceutical companies can more closely align research and development priorities with market realities. This new approach will be challenging, and even a little frustrating, because payer priorities change over time. But pharmaceutical companies must listen, respond, and evolve based on what their customers are saying.

At all of these stages, businesses need a guide who understands these issues and can work in partnership with them to develop effective solutions to the challenges.

How do you know if you are Innovating Effectively?

Innovation is not for the risk averse. For example, in the context of the Life Sciences environment, there are a lot of moving parts connected to effective and efficient research and development delivery. Emphasis on quality of care, adhesion to regulation, increasingly challenging reimbursement policy, and the risk of litigation heighten the stakes relative to introducing change in the medical environment. How can cutting edge technology successfully migrate into healthcare without also being a risky proposition? How is the landscape changing? Or, more importantly, how can you find out how it is changing?

The answer is preparation, iteration, and prototyping as driving components of the development process. This discussion is beyond the scope of this text, but according to the experiences I have had in the research and development and new product development arena over the past seventeen years, I recommend focusing on methods and strategies for proactively identifying and responding to obstacles to success relative to technology adoption, including:

1. Extracting user needs beyond Voice of the Customer (VOC).
2. Utilizing effective co-development with high value-add supply chain partners.
3. Effectively identifying and speaking to ALL of your stakeholders .
4. Making sure your requirements are the RIGHT requirements.
5. Having the power of an effective and efficient process of development optimizing across the most critical variables, which could require substantial iteration in the development process.

16. Conclusion

Competitive markets and demanding customers require updated and 'refreshed' products and services. Even small changes to products and services can have an impact on competitiveness. Markets are also becoming more fragmented, requiring product and service variants developed specifically for custom market needs. Simultaneously, technologies are affording researchers increased opportunities for their exploitation within novel products and services. It is critical to appreciate that the successful development of products and services is inextricably intertwined with efficient and effective and ideally optimized development of processes that produce them. Product and service development success is governed by successful and efficient research and development processes. In order to achieve effective and efficient research and development and product innovation it is not uncommon for firms to use cross-functional teams, together with teams focused exclusively on a particular product differentiation effort.

Establishing the sense of connectivity for scientists and engineers with the overall business strategy is absolutely critical to achieve efficiency and effectiveness in the research and development efforts of a firm. Scientists who are connected to the business strategy will have a better clarity as to what to do in certain situations and create a valuable sense of ownership that the employees value. Clear transparency of business strategy shared across the organization will create enhanced ownership, loyalty, and creativity by most employees. Translating the overall business unit or plant objectives into goals and objectives at the lower levels in critical to create that sense of transparency and ownership. Clear and open communication both from the business leadership down and from the lower level scientists up to senior management will quickly facilitate the understanding of problems, constraints, and new opportunities and catalyze strategic innovation.

There is no single model that companies utilize in new product development; in fact, there are many. Nevertheless, these share some common denominators including concept generation, concept screening, preliminary design, design evaluations and improvement, prototyping and final design, and developing the efficient and effective operations process to deliver successful outcomes.

A visionary research and development, or new product or service development, that is not linked early on to excellent operational and governance processes, cannot be implemented. Conversely, operational excellence may lower costs, improve quality, and reduce process and lead times, but without a strategic research and development vision and guidance, it is unlikely to enjoy sustainable success from its operational improvements alone. High performance operating processes in research and development are critical, and when combined with proactive implementation and governance, will result in successful and strategic research and development that could serve as a sustaining advantage to companies in the near and long term.

Most companies face constraints on the resources at their disposal to innovate and conduct research and development; therefore, they need to allocate these resources selectively to achieve the outcomes they seek. We have introduced linear programming in this chapter; it is a valuable tool in tackling problems involving resource constraints.

This chapter has discussed several of the benefits of developing and nurturing a strategic research and development strategy. The benefits are not only external, in gaining the upper hand on the competition, but also internal. Internally, employees demonstrate greater loyalty, creativity, and a greater understanding of what the company is trying to achieve. If some of the tools and methodologies discussed in this chapter are placed into practice, management will find that these continue to fuel further strategic innovative behavior and that the overall benefit is compounded over time.

17. References

An Introduction to Linear Programming, Steven J. Miller, March 31, 2007, Mathematics Department, Brown University, 151 Thayer Street, Providence, RI 02912

Cogliandro, J. (2007) *Intelligent Innovation*, J.Ross Publishing, Fort Lauderdale, FL.

Gibson, R. and Skarzynski, P. (2008), *Innovation to the Core*, Harvard Business Publishing, Boston, MA.

Holtzman, Yair, *Innovation in research and development: tool of strategic growth*, Journal of Management Development p. 1037-1052, JMD Vol. 27, Number 10, 2008

Mandelbaum, A. (1996), *"Getting the most out of your product development process"*, Harvard Business Review, March-April.

Mandelbaum, A. (1996), *"Getting the most out of your product development process"*, Harvard Business Review, March-April.

Mobilizing an R&D Organization Through Strategy Cascading by Christoph H. Loch, 2008, Industrial Research Institute, Inc.

The Journal of Product Innovation Management (2002) *Implementing a strategy-driven performance measurement system for an applied research group*, Christoph H. Loch, U.A, Staffan Tapper, INSEAD, Fontainebleau, France & Witts Graduate School of Business, Johannesburg, South Africa

3

Economic Impact of the Adoption of Enterprise Resource Planning Systems: A Theoretical Framework

Wai-Ching Poon, Jayantha Rajapakse and Eu-Gene Siew
Monash University Sunway Campus
Malaysia

1. Introduction

According to Deloitte Consulting, Enterprise Resource Planning (ERP) systems are packaged business application software suites that allow an organization to automate and integrate the majority of its business processes, share common data and practices across the entire enterprise, and produce and access information in a real-time environment. The scope of an ERP solution includes financials, human resources, operations logistics, sales and marketing modules (Ragowsky & Somers 2002).

The benefits that ERP brings to organization are multidimensional and include tangible and intangible benefits (Shang & Seddon, 2002). One of the key characteristics of ERP systems is the potential for data and process integration across different units of an organization (Deloitte, 1999; Ross & Vitale, 2000; Markus et al., 2000; Volkoff et al., 2005). Such integration enables real-time decision-making based on ready access to reliable up-to-date information. ERP also allows centralization of data and streamlining of business process. This results in efficiency of business process and reduction in cost (Spathis & Constantinides, 2004). Many studies have shown the benefits of ERPs, ranging from improving productivity (Hitt et al., 2002; Ifinedo & Nahar, 2006), decision support benefits (Holsapple & Sena, 2005) and integration benefits of various information systems (Hsu & Chen, 2004).

According to Huang et al. (2004), ERP generates tremendous amount of information goods, helps create value chain and increases value-added activities by categorizing available information, such as information about customers, suppliers, transactions cost, and the price of unit sold. Information could be categorized according to cost-benefit information with respect to logistic and shipping, marketing, sales and purchasing, and resource allocation, after sale service support, and resource optimization. A few assumptions are requisite. For example, the assumptions of constant returns to scale and perfect competitive in the product market are often imposed in the estimation of input shares. Constant returns to scale refer to the output increases at the same rate as the inputs in the production function. To build up information goods, there involves high level of fixed cost and these cost of productions remain constant in the future. That is to say, all information goods are replicated with zero or very low marginal cost. With information goods in the marketplace, firms with ERP systems are able to gain competitive advantage, a more practical coordination and interaction between supplier-customer and hence minimize cost and ultimately optimize market efficiency.

In this paper, to further extend the above study, we propose an economic analysis framework of the impact of ERP at the firm level. We will use economic production theory to examine ERP role in this regard. This is important because it extends the understanding of ERP system's impact and this framework can be used as a basis for research.

This paper is organized as follows: Section 2 describes related research in this area, Section 3 presents the conceptual framework, Section 4 provides a brief discussion and Section 5 concludes the chapter.

2. Related works

Many research studies examine the relationship between IT and economic performance or productivity growth. These studies on economic impact are on firm level, sub-plant level, and country level. Productivity is the elementary economic measure of a technology contribution. There has been considerable debate whether information technology (IT) revolution was paying off in higher productivity (Dedrick et al., 2003). However, results are inconclusive. The Nobel Laureate economist Robert Solow said that "we see computers everywhere except in the productivity statistics" (cited in Brynjolfsson, 1993). Prior to 1990s studies found productivity paradox between IT investments and productivity in the U.S. economy. Thereafter, many studies found greater IT investment and revolution observed in higher productivity gains at both firm and country levels. However, there were studies who found IT capital has marginal impact on technical progress (Morrison & Siegel, 1997), and some claimed that IT has insignificant contribution to output growth (Oliner & Sichel, 1994; Loveman, 1988).

Four explanations for this productivity paradox include mis-measurement of outputs and inputs, lag effects as a result of adjustment, relocation and rakishness of profits, and mismanagement of information and technology. According to learning-by-using model, the optimal investment strategy sets marginal benefits lesser than marginal costs in the short run. But firm will only see the impacts after sometime due to lag effect and increasing economies of scale might only be experienced in the long run. Kiley (1999, 2001) argues that adjustment costs have contributed to some negative relationship between IT and productivity and he further argued that adjustment costs have created frictions that cause investment in IT capital to be negatively associated with productivity. Meanwhile Roach (1998) argues that much of the productivity stimulation is due to the secular trend toward service related industries that are caused by rising mis-measurement errors, such as over-allow work flexibility, causing unnecessary longer overtime labor hours claims. Thus, actual labor hours in the IT related industries may not be reflecting the true productivity growth figure.

Some studies have analyzed firm level data and find evidence of significant and positive returns from IT capital investment (e.g., Brynjolfsson & Hitt, 1996; Dewan & Min, 1997). The advantage of the firm level approach is that it gives better measurement of IT contributions to both quality and variety of products that covered at aggregate level. Some others have examined economy level time series data to quantify the contribution of IT toward output growth of a single country, with mixed findings on the contributions of IT.

The above discussion is on IT in general. Now turning to ERP as a specialized area of IT, from the extant literatures, many studies examine the relationship between ERP systems and its economic impact. Research on the impact of ERP can be broadly divided into level of analyses (for example, firm level and sub-plant level) or the different dimensions of impact (for example, financial, operational and managerial). Studies on firm level focus on the effects on the whole organization. These can be financial impacts, or the five classification by Shang & Seddon (2002); namely operational benefits, managerial benefits, strategic benefits, IT infrastructure benefits, and organizational benefits.

Studies on financial impacts of ERP typically measure performance of financial statements (Poston & Grabski, 2001), financial ratios (Hendricks et al., 2007; Hunton et al., 2003; Matolcsy et al., 2005; Poston & Grabski, 2001; Wieder et al., 2006; Wier et al., 2007) and share price of the company (Hendricks et al., 2007; Hitt et al., 2002). These performances are usually compared for a group of companies that adopted ERP against those companies that do not over a period.

Results from these research listed above have consistently indicated that financial performance will be negatively affected in the first two to three years during the ERP implementation and only after two to three years, will the firm see improvements (Hendricks et al., 2007; Hitt et al., 2002; Hunton et al., 2003; Matolcsy et al., 2005; Poston & Grabski, 2001; Wier et al., 2007). From the list above, only one study that seem to contradict the claim that there is no significant differences between adopters and non-adopters (Wieder et al., 2006). However, that study did not account for the time after the ERP implementation has taken place and the small sample size.

Besides the financial impact, it was found that the benefits of implementing ERP systems extend to the operational (Cotteleer & Bendoly, 2006), managerial, strategic and planning and control process integration of supply chain management (Su & Yang, 2010). Managerial, operational and IT infrastructure benefits was observed one year after implementation of ERP (Spathis & Ananiadis, 2005). ERP was also shown to improve the accounting process (Spathis & Constantinides, 2004).

Research into sub-plant level found that the benefits of ERP is more when the sub-units ("business function or location") are more dependent on each other and less when the sub-units are vastly different (Gattiker & Goodhue, 2005). Analysis and research of the impact of ERP at the firm aggregate level has been scarce although there are many similar IT research at this level. Huang's (2004) economic analysis of ERP as information goods generated positive externalities value which will increase as more numbers of suppliers and customers of the firms are interconnected. This has been called the network effect. The authors also argue that although the cost of implementation of ERP is high but the cost supplying information is almost zero once the adoption of ERP system is on.

3. Conceptual framework

The purpose of this section is to explain and justify the conceptual framework proposed by the authors. The framework is based on a synthesis of the economic production theory and network externalities. In other words, the framework classifies economic impact based on a productivity function and the network externalities.

How inputs are transformed to output is commonly illustrated in a production function. As seen in the Section 2 many studies examine the effect of ERP on productivity growth by examining stock prices and profitability. The more recent studies use panel analysis and the longitudinal approach to estimate inputs to Gross Domestic Product (GDP) outputs and its returns from IT investment in the aggregate level. Generally, output growth in firms, sectoral and the country level may be due to an increase in input level, improvement in the quality of input, and productivity growth of inputs. Furthermore, the effect of IT adoption in a neoclassical theory rests on labor productivity and can be explained using capital deepening effects (Stiroh, 1998; Jorgenson & Stiroh, 1999), embodied technological change, and productivity spillovers. Capital deepening refers to the growth of capital (e.g. information processing equipment and software) that workers have available for use in a firm. ERP systems may allow total factor productivity gains since it allows production of improved capital goods at lower prices via some production spillovers or positive externalities effects (see Bresnahan, 1986; Redmond, 1991; Bartelsman et al., 1994).

A positive network externality has been widely used in the study of technology adoption. It is an economic concept describing a consumer's demand may be affected by other people who have purchased the good, and gained the benefit in consumption due to the widespread adoption of physical goods and services. Earlier studies (eg. Jensen, 1982) on internet and e-commerce have shown that people are more likely to adopt certain technology if others within the same industry or region likewise use it. An ERP adopting organization can integrate the ERP system with its suppliers and customers thereby creating an electronic market. The ERP that enables electronic markets comprised of supply and demand networks to facilitate information exchange (Huang et al., 2004). The suppliers and customers may or may not be using ERP systems. However, they can access the information goods generated by the ERP. Thus, ERP in the electronic markets serves as the information processing function to generate and exchange information among suppliers and customers. This electronic transfer of information goods can reduce the cost of paperwork and processing requirements of all the parties involved. Hence, marketable information goods produced by ERP would bring additional profits to organization. Next section, we discuss the production theory and network effects respectively in detail. The proposed framework is depicted in Figure 1 below.

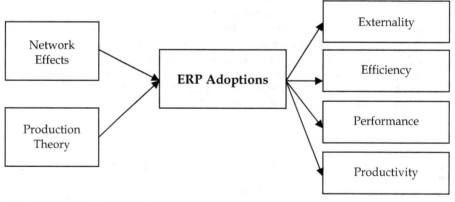

Fig. 1. The proposed framework.

3.1 Network effects

Network effects impact technology choice (Katz & Shapiro, 1994). Network effects arise when there is interdependence between different components of an economic system (Young, 1928). We may ask questions such as how does a change in technology affect the increase in output and will this become an incentive for firms to exploit the increasing returns for adopting this technology (Arthur, 1996). Integrated with e-data interchange, ERP can be used to restructure supply chain operations via B2B e-hubs with supply chain partners to run transactions in real time (Zeng & Pathak, 2003). As more supply-chain partners become integrated with the ERP systems, the entire supply chain can be integrated and streamlined with other functions to be more competitive, reduce the marginal cost of productions, increase the profitability of the organization, and maximize productivity of the firm. To enable electronic markets, internal networks structures are important fundamental economic characteristic.

According to Majumdar and Venkataraman (1998), there are three network effects in the literature. The first is conversion effect, driven by operations-related increasing returns to scale that firms enjoy in converting from one system to another. The second is consumption effect, driven by demand-side increasing returns to scale that it is a firm-level effect that arises where customers are interconnected. The third is an imitative effect that arises when the inter-firm information flows are induced by imitation pressures between firms.

The conversion effect arises when there are increasing returns moving towards the usage of advanced technology. Cost-benefit analysis hypothesizes that inputs affect outputs to determine the identifying statements of organization goal such as maximization revenue, minimize cost, and maximize profits. An initial ERP adoption is likely to involve high cost. There are incentives to convert to the new technologies because of the possibilities of enhancing operating efficiencies. The greater the relative size, the higher the incentive to exploit conversion effects since there are larger numbers of customers and suppliers who provide the means to write-off adoption costs.

Consumption effect exists when there is demand interdependence among customers. This effect is enhanced by the density and composition of customers in the network. When there is high network density and variety of user population in a network, there will be an increase in network functionality. This implies a larger potential market, and therefore brings about higher utility to the customers. Hence, network density and user population are expected to be positive at all times. Meanwhile imitation effect is salient in industries where firms share a common infrastructure, and that many channels are available for dissemination of information between those interconnectivity firms and the nature of equipment. Therefore there are increasing returns to the inter-firm spread of information (Markus, 1992). When managers face a new technology with uncertain trade-offs, imitation provides a solution with low risk (Majumbar & Venkataraman, 1998). Therefore, the imitative effect will have positive effect on the new technology, the ERP system adoption, at all times.

3.2 Impact of network externality on the adoption of ERP

There are many models to test for the presence of network externalities on the adoption of ERP process (Katz & Shapiro, 1986; Farrell & Saloner, 1986; Cabral, 1990). For instance,

Cabral's (1990) model allows for heterogeneity in the benefits available from network dynamic. The benefits from membership upon adoption are B(h,n,t), where n is the measure of adopters at time t, h is a parameter that characterizes a technology (the higher the h for a firm, the higher is the benefit from adopting ERP membership, all other things remain equal), and t is time. The assumption that there are externalities in network participation is captured by Bn > 0, Bh > 0, and Bt > 0. The latter assumption reflects the exogenous trend to increase benefits from adopting the shared network technology, reflecting improvements in the ERP technology itself.

Since information can be reproduced at zero or very low marginal cost, and supply chain network using ERP system can be connected in constant returns to scale, all inventory information can be stored in the system and causing information supply networks to exhibit positive network externalities of production. Market dynamic works in such a way that the supply curve with network externalities of production starts high and decreases toward zero. The impact of network technologies on financial institutions depends on assets, number of employees, and number of branches (Zhu et al., 2004).

Positive network externalities of consumption are a kind of demand side network economics of scale. It is highly dependent on the number of organization already connected to the ERP systems. If there are large numbers of organization connected to ERP systems, the willingness to pay for the marginal organization is also low because every organization that valued it higher has already connected to ERP systems. Therefore, an organization's demand for the information goods depends on the marginal willingness to pay. The reservation price for information goods is determined by the marginal willingness to pay, which at first increases and then decreases with the number of organizations connected to the demand network (Huang et al., 2004). Therefore, the demand curve for information goods with network externalities of consumption is hump-shaped. Hence, for market dynamics, the supply and demand curves with network externalities will intersect only if there is a small number of organization connected to the markets and information good exchange are low, i.e., happen when there is a low equilibrium level (Majumdar & Venkataraman, 1998).

3.3 Economic production theory

Economic evaluation orientation to IT impact ranges from relatively simple cost-benefit analysis (King & Schrems, 1978) to rigorous production function (Kriebel & Raviv, 1980) that mostly focuses on profit of the organization. Mapping major microeconomic production indicates that ERP has been used in operational or management control decisions for production modeling. ERP systems have been used in diverse areas of transaction processing in accounting, finance, marketing and management.

The production function is a commonly use tool in analyzing the process of economic growth and performance of a firm. A production function relates the inputs of the production process. A firm production function uses decisions and firm resources (e.g. labor, raw materials, information, IT capital, non-IT capital, decisions, inventory decision, and etc) as inputs and the attainment of organization goals (eg. profit maximization, sale maximization, revenue maximization, or cost minimization) as output to achieve economic performance outcomes (eg. economic growth, labor productivity, profitability, or overall welfare). A productive firm will generally enjoy higher profitability, or a firm is perceived to be productive if a firm is able to produce the same output level with fewer inputs and thus experiences a cost advantage, or produces higher quality output with the same level of inputs and enjoys a price premium.

Many scholars have examined the relationship between IT-economic performance or IT-productivity growth. Input productivity is important determinants of economic growth. Productivity is a measure of how efficient resources are converted into goods and services in a production process. It can be calculated as the ratio of output to input. Hence labor productivity is the output produced per unit of labor, and it can be calculated using total output divided by the total unit of labor employed. Labor productivity always means average product of labor or average productivity. Therefore, average productivity (AP) is calculated by output/labor input, and it is often used as a measure of efficiency. When a firm experiences productivity increases, it means that output per unit of labor input has been increased. However, as more and more of one input (eg. labor) is added with a given amount of another input (eg. capital), the increases in output will eventually decline. This is called the law of diminishing returns. Similarly, as worker acquires more capital, there is diminishing return to that capital. If this process continues in a longer period, the growth will gradually slow to zero.

Total factor productivity pertains to the efficiency of the inputs mix to produce output. Efficiency gains could be achieved through more effective distribution arrangements, greater economies of scale, better management, shift from low productivity production to high productivity activities, the adoption of new technology, innovation and intervention, or the replacement of old capital, or retrained the workers that enable greater output production using the same level of input mix. There are generally two factors that affect productivity. The first is human capital and the second is technology. Human capital refers to worker's investment in education and training that could upgrade the skills of the existing labor force and improve the quality of labor force, with more IT literate and more congenial staff, they are able to easily adapting newly installed technologies, and the increase in human capital investment is a major contributor to the long-run economic growth. This is also called the embodied technical progress. Meanwhile, investment in technology involves the way inputs are mixed in the firm, such as innovation and invention of new products, improvements in organize production, advances in management and industrial organization, and better manage economic factors of productions that increase the output level even when the amount of labor and capital are fixed. Adoption of ERP systems can produce all such benefits as identified in extant literatures, such as Shang and Seddon (2002), Huang et al. (2004), and Wieder et al. (2006). This is also called disembodied technical progress. The productivity gain resulting from technological progress seem unlikely to be sustainable over the very long run whenever we reach the point of diminishing returns to the technology investment (Sharp et al., 2006). In terms of ERP systems, it is necessary to upgrade the system quite frequently to keep up with the technological and business changes. Such upgrades require new capital infusions.

3.4 Model

Economic theory shows that the basic way to measure productivity is the standard firm production model that is based on a gross output production function that relates firm gross output to the factors of productions such as capital and labor, intermediate inputs such as energy and raw materials, and total factor productivity. The simple model of production shows the relationship between inputs and outputs is formalized by a simple production function as:

$$Y_t = f(K_i, L_i, M_i, ...) \tag{1}$$

where Y represents the firm's output or return on assets (ROA) or return on sales (ROS) (Wagner et al., 2002) during a period, K denotes the capital usage during the period, L represents hours of labor work, M represents raw materials used, and notation represents the possibility of other variables influencing the production process.

The same level of output can be produced with fewer inputs. For example, with a level of capital input of K, it previously took L_2 unit of labors to produce Y_0, now it takes only L_1. Output per worker has risen from Y_0/L_2 to Y_0/L_1. However, it is noteworthy that an increase in capital input to K_2 could also lead to a reduction in labor input to L_1 and produce similar level of Y_0. If this is the case, output per labor would also rise, but there could have been no technical progress. To measure technical progress we could write in a simple equation as follows:

$$Y = Z(t)f(K, L) \tag{2}$$

where the term $Z(t)$ represents technical progress as a function of time that shows the factors that determine Y other than K (capital hours) and L(labor hours). Technical progress in the Cobb-Douglas production function could be represented by $Y = Z(t)f(K, L) = Z(t)K^\alpha L^{1-\alpha}$, for simplicity, we assume constant returns to scale and that technical progress occurs at a constant exponential mode (θ_t). We can rewrite the function as: $Y = Z(t)f(K, L) = Z(e^{\theta t})K^\alpha L^{1-\alpha}$.

Suppose that $Z = 10$, $\theta = 0.01$, $\alpha = 0.5$, and the firm uses input mix of 2 units of capitals and labors each (K=L=2) currently (at time t=0), therefore output is 20 ($Y_t = 10e^{0.01(0)}2^{0.5}.2^{0.5}$). After 10 years, the production function with this input mix becomes 22 ($Y_{t+10} = 10e^{0.01(10)}2^{0.5}.2^{0.5}$). However, if output increases more rapidly than the inputs, given the fixed technology, this would imply that there is an increasing returns to scale. With the adoption of ERP systems, it is believed that technical innovation operates through the positive effects. These positive externalities help to generate increasing returns to scale and drive the firm's performance.

To account for total factor productivity or multifactor productivity, term Z is included in the function. They can be represented in a function as: $Y_t = f(K_i, L_i, M_i, Z_i)$

where Y is real output or ROA or ROS, K is capital, L is hours worked, M is intermediate inputs or raw material used, and Z is a total factor productivity index for firm i.

Generally, we perceive competitive market structure exists in capital and labor, therefore constant return to scale is assumed. We can rewrite the growth rate of real output equals to the growth rates of the capital and labor inputs weighted by their shares in real gross output as follows:

$$\omega(Y_t) = \omega(K_i)W(K_i) + \omega(L_i)W(L_i) + Z_i \tag{3}$$

where $\omega(Y_t)$ is the growth rate of output, ROA or ROS, $\omega(K_i)$ is the growth rate of capital investment (including net depreciation), $\omega(L_i)$ is the growth rate of labor, and W(K) and W(L) are the weighted shares of capital and labor in the firm, respectively. $\omega(K_i)W(K_i)$ is the growth rate of capital multiplied by the ratio of capital to labor, which we called as marginal product of capital. Similarly $\omega(L_i)W(L_i)$ is the growth rate of labor multiplied by the ratio of labor to capital, which we called as marginal product of labor, and Z_i is the productivity efficiency factor, which is a residual term that is not accounted for by the growth of labor and capital.

Suppose that a firm has a growth rate of output of 5 per cent, the growth rates of capital and labor of 10 and 2 percent, respectively, and the weighted shares of capital and labor are 20 and 80 percent, respectively. Therefore Z_i has to be equaled to 0.014. This reflects that technical progress account for slightly less than 1.5 percent of the output growth of 5 percent.

$$\omega(Y_t) = \omega(K_i)W(K_i) + \omega(L_i)W(L_i) + Z_i$$

$$0.05 = 0.2(0.10) + 0.8(0.02) + Zi$$

$$Z_i = 0.014$$

Past studies present econometric estimates using Cobb-Douglas production function (e.g. Gera et al., 1999; Brynjolfsson & Hitt, 1996; Lehr & Lichtenberg, 1998), cost function (e.g. Morrison & Siegel, 1997) or panel estimation (e.g. Stiroh, 2001). There are some microeconomic productions properties apply to the Cobb-Douglas production model for ERP systems, assuming a constant elasticity of substitution (CES). The CES production technology exhibits a constant percentage change in factor (e.g. capital and labor) proportions due to a percentage change in marginal rate of technical substitution (MRTS). MRTS is the amount of one input that must be substituted for one unit of another input to maintain a constant level of output. First is marginal productivity. It is the rate of increase of the output for a small increase in the input. The Law of Diminishing Marginal Productivity will set in if the marginal product is positive but diminishing. Second is input substitutability, where inputs will be substituted more of one input and less of another to produce the same level of output. Third, it is assumed that decision making is in steady state (i.e., constant input and output levels, all other parameters remain unchanged).

Alternatively, one can study how different types of capital affect labor productivity growth. This can also be carried out using Cobb-Douglas production function that can explicitly decompose capital into IT-related and non-IT related categories. This can be written using Cobb-Douglas production function in the form of $Y_{it} = f(IT_i, K_{it}, L_{it})$. It can be tested for many firms, with i = 1, 2, ..., N using years of data, in Year t = 1, 2, . . ., T. The output production Y_{it} is annual performance of the firm, and the inputs are IT capital stock (IT_{it}), non-IT capital stock (K_{it}) and annual labor hours employed (L_{it}). For example, for a data of 20 firms over the period of 10 years, then N=20, T=10. Normally, the regression model will be controlled for firm effect and year specific effect. For the functional form of f(.), we can write the Cobb-Douglas production function at the log form as follows:

$$\log Y_{it} = \alpha + \sigma_t + \beta_{IT} \log IT_{it} + \beta_K \log K_{it} + \beta_L \log L_{it} + V_i + e_{it}, \qquad (4)$$

where σ_t is a time effect captured by year dummy variables in the regression, V_i is a firm-specific effect invariant over time, and e_{it}, is the random error term in the equation, representing the net influence of all unmeasured factors (Dewan & Kraemer, 2000). From this Cobb-Douglas function, the output elasticities of β_{IT}, β_K, and β_L that measure the increase in output associated with a small increase in the corresponding inputs could be estimated. For example, the output elasticity of IT capital (β_{IT}) shows the average percentage increase in GDP for a 1% increase in IT capital. In other words, it is the output elasticity of IT capital. Other output elasticity parameters with respect to capital and labor have analogous interpretations.

Pooling data from firms increases the variation in the variables, and is therefore crucial to account for firm effects. There are two general models to capture cross-sectional heterogeneity. They are fixed effects and random effects models. Fixed effects approach could be carried out by putting in dummy variables. This is very costly since we can easily losing the degrees of freedom. This makes the random effects model more engaging. However, the random effects model requires the potentially restrictive assumption that the V_i be uncorrelated with the regressors to avoid inconsistency (Greene, 1990).

In practice, it is not easy to get good proxy for capital stock. Among those proxy measure capital stock in total factor productivity are the rate of R&D investment, and rate of investment in computers and investment in human capital (Siegel, 1997), the number of information systems workers (Brynjolfsson & Hitt, 1996), investment figures to measure the increment to capital (Dowling & Valenzuela, 2004), return on capital employed (Wagner et al., 2002), and inventories reductions to show a higher efficiency of producing and delivering goods (Varian et al., 2002). They have reported that an increase in investment in IT has a positive effect on the productivity performance in a given firm. Meanwhile, indicators commonly used as proxy for human capital includes total years of schoolings derived from educational enrolment ratios, international test scores, number of workers with tertiary education (Barro & Lee, 2001), the number of educational years in higher education and the experience of the works (Barros et al., 2011), labor in terms of man-hours, man-years worked, labor cost as a fraction of profit (Dewan and Kraemer, 2000), or construct a series by multiplying the labor series by an index to show rising educational attainment over time, or by introducing a new factor of production, such as education and training, and then measure its contribution to output separately (Dowling & Valenzuela, 2004). Bresnahan et al. (2002) use IT demand, human capital investment and value-added as dependent variables, and they found that IT, organization change and human capital, technological and organization changes are complimentary to each other, and these variables can boost up market value of firms.

From the empirical studies, we propose that the output for production function can be measured as follows:

$$Y_t = f (K_i, L_i, H_i, M_i, Z_i) \tag{5}$$

For the functional form of f(.), we can write the Cobb-Douglas production function at the log form as follows:

$$\log Y_{it} = \alpha + \beta_K \log K_{it} + \beta_L \log L_{it} + \beta_H \log H_{it} + \beta_M \log M_{it} + \varphi_{Tt} DT_t + Z_i + e_{it}, \tag{6}$$

where subscript i is the ith firm and t is the time period; the output Y_{it} is annual performance of the firm, or output production, or ROA or ROS, and the inputs are physical capital stock (K_i), human capital variables expressed in average number of employees with tertiary education (H_i), annual labor hours employed (L_i), M is intermediate inputs or raw material used, DT_t is the dummy variable for different years to capture technology change, and the parameter φ_{Tt} can be used to measure technical level over time. The technical progress or the rate of change in technical level can be calculated using $\varphi_{Tt} - \varphi_{Tt-1}$. Z_{it} is the random errors, reflecting total factor productivity for firm, with $Z_{it} \sim N(0,\sigma^2)$ and e_{it} is a non-negative truncated normal random error with the probability distribution of $e_{it} \sim N(\mu, \sigma^2)$.

4. Discussion

How much of an economic transformation is the ERP likely to produce in an organization? How will the ERP systems affect the performance of the organization and the skills of the people? How customizing ERP information affects market dynamic? Will it be significant determinant in sustaining and maintaining the dramatic increase in productivity recorded since the mid-1990s?

The economic contributions of information technology in general and ERP in particular, have important policy implications and have attracted the attention of researchers (Dewan & Kraemer, 2000). Cost saving and productivity have been reported positive relation in computer and software industries (Gordon, 2000; Oliner & Sichel, 2000). The cost savings are largely projected to be one-time savings for each firm or spread over in individual firm, while at the sector level a process of diffusion from first-adopters to followers should generate a pattern of productivity savings (Litan & Rivlin, 2001). Cost savings from the networking of ERP system often offer some important gain to consumers from added convenience, variety of product mix, and customization that ERP makes possible. These significant savings could be generated from the large productivity increases in ERP adoption.

Performance of an organization could be improved through better management, innovation and re-skilling of the workforce. ERP forces firms to conform and standardize their business process to the best practices. Thus, ERP systems innovate the old business processes and thereby make the processes more efficient. As best practices streamline the business processes the management of an organization could make better decisions in terms of meeting market demands such as introduction of new product lines etc

ERP adoptions facilitate integration of divisions within a firm as well as externally with suppliers and customers. Such externalities offer the benefit of connecting and communicating between different systems, not having to maintain separate systems and ability to easily share information between systems. This would result in better strategic and operational decision making and thus, higher profits. For example, in the petrochemicals industry, it is difficult to find companies without ERP because sharing of information electronically is crucial for their survival (Davenport, 1998).

The ERP-skilled workforce can improve the firm's performance in several ways. For example, cycle time reduction through completing tasks with less time, proposing continuous improvements to the business processes etc. The benefits of added convenience and customization are inherently much more difficult to quantify (Varian et al., 2002), and are not likely to show in GDP.

5. Conclusion

ERP technology enables firms to cut costs, improve transactions and enlarge markets, foster productivity growth, and improve the skills of the workforce. Firms that discover ways to use the ERP productively will be on the cutting edge of their markets. To conceptualize these impacts, we proposed a framework based on the production theory and the network effects of the information goods. In the production theory, ERP system is treated as a capital investment. The premise is that the higher the investment in capital the higher the productivity. Higher productivity results in higher output of goods and services per unit of raw materials. As mentioned earlier, these productivity increases result from better integration effects (network effects), cost savings from ERP and the streamlined of business processes.

Besides the internal impact from ERP that arise from the production function, we also incorporate the external impact from the network effects. As more firms adopt ERP systems within a supply chain of a firm, the benefits that these firms bring to new firms adopting ERP are increasing to scale. This is termed the network effects and can comprise of the conversion effect, consumption effect and imitative effect.

The proposed framework explains why firms adopt ERP despite the risks and costs; how it impacts internally through the production function and how the external factors through the network effects encourage firms to adopt ERP. Likewise, this framework allows an understanding of how ERP affects the firm as a whole from the production function and network effect.

6. References

Arthur, W.B. (1996). Increasing returns and the new world of business. *Harvard Business Review*, Vol.74, No.4, pp.100-109.

Barro, R.J., & Lee, J.W. (2001). International data on educational attainment: updates and implications. *Oxford Economics Paper*, Vol.3, pp.541-563.

Barros, C.P., Guironnet, J.P., & Peypoch, N. (2011). Productivity growth and biased technical change in French higher education. *Economic Modelling*, Vol.28, pp.641-646.

Bartelsman, E.J., Caballero, R.J. & Lyons, R.K. (1994). Customer- and Supplier-Driven Externalities. *American Economic Review*, Vol.84, No.4, pp.1075-1084.

Bresnahan, T.F. (1986). Measuring the Spillovers from Technical Advance: Mainframe Computers in Financial Services. *American Economic Review*, Vol.76, No.4, pp.741-755.

Bresnahan, T.F., Brynjolfsson, E., & Hitt, L.M. (2002). Information technology, workplace organization, and the demand for skilled labor: Firm level evidence. *Quarterly Journal of Economics*, Vol.117, pp.339-376.

Brynjolfsson, E. (1993). The productivity paradox of information technology. *Communications of the ACM*, Vol.36, No.12, pp.67-77.

Brynjolfsson, E. & Hitt, L.M. (1995). Information technology as a factor of production: The role of differences among firms. *Economics of Innovation and New Technology*, Vol.3, No.3-4, pp.183-200.

Brynjolfsson, E. & Hitt, L.M. (1996). Paradox lost: Firm level evidence on returns to information systems spending. *Management Science*, Vol.42, No.4, pp.541-558.

Cabral, L.M.B. (1990). On the Adoption of Innovations with 'Network' Externalities. *Mathematical Social Sciences*, Vol.19, pp.299-308.

Cotteleer, Mark J., & Bendoly, Elliot (2006). Order lead-time improvement following enterprise information technology implementation: an empirical study. *MIS Quarterly*, Vol.30, No.3, pp.643-660.

Davenport, T. H. (1998). Putting the enterprise into the enterprise system. *Harvard Business. Review*, Vol.76, No.4, pp.121-131.

Dedrick, J., Gurbazani, V. & Kraemer, K.L. (2003). Information technology and economic performance: A critical review of the empirical evidence. *ACM Computing Surveys*, Vol.35, No.1, pp.1-28.

Deloitte Consulting (1999). *ERP's Second Wave: Maximizing the Value of ERP-Enabled Processes*. New York: Deloitte Consulting.

Dewan, S. & Kraemer, K. (2000). Information technology and productivity: evidence from country-level data. *Management Science*, Vol.46, No.4, pp.548-562.

Dewan, S. & Min, C.K. (1997). Substitution of information technology for other factors of production: A firm level analysis. *Management Science*. Vol.43, No.12,pp.1660–1675.

Dowling, J.M. & Valenzuela, M.R. (2004). Economic Development in ASIA. Thomson Learning, Singapore.

Farrell, J. & Saloner, G. (1986). Installed Base and Compatibility: Innovation, Product Preannouncements, and Predation. *American Economic Review*, Vol.76, No.5, pp.940-955.

Gattiker, T. F., & Goodhue, D. L. (2005). What happens after ERP implementation: understanding the impact of inter-dependence and differentiation on plant-level outcomes *MIS Quarterly*, Vol.29, No.3, pp.559-585.

Gera, S., Gu, W. & Lee, F. (1999). Information technology and labour productivity growth: An empirical analysis for Canada and the United States. *Canadian Journal of Economics*, Vol.32, pp.384-407.

Gordon, R. (2000). Does the "new economy" measure up to the great inventions of the past? *Journal of Economic Perspective*, Vol.14, No.4, pp.49–76.

Greene, W. H. (1990). *Econometric Analysis*. Macmillan Publishing Company, New York.

Hendricks, K., Singhal, V. & Stratman, J. (2007). The impact of enterprise systems on corporate performance: A study of ERP, SCM, and CRM system implementations. *Journal of Operations Management*, Vol.25, No.1, pp.65-82.

Hitt, L.M. & Brynjolfsson, E. (1996). Productivity, business profitability, and consumer surplus: Three different measures of information technology value. *MIS Quarterly*, Vol.20, No.2, pp.121–142.

Hitt, L. M., Wu, D. J. & Xiaoge, Z. (2002). Investment in Enterprise Resource Planning: Business Impact and Productivity Measures. *Journal of Management Information Systems*, Vol.19, No.1, pp.71-98.

Holsapple, C. W. & Sena, M. P. (2005). ERP plans and decision-support benefits. *Decision Support Systems*, Vol.38, pp.575-590.

Hsu, L. L. & Chen, M. (2004). Impacts of ERP systems on the integrated-interaction performance of manufacturing and marketing. *Industrial Management & Data Systems*, Vol.104, pp.42-55.

Huang, M.H., Wang, J.C., Yu, S. & Chiu, C.C. (2004). Value-added ERP information into information goods: an economic analysis. *Industrial Management & Data Systems*, Vol.104, No.8, pp.689-697.

Hunton, J. E., Lippincott, B. & Reck, J. L. (2003). Enterprise resource planning systems: comparing firm performance of adopters and nonadopters. *International Journal of Accounting Information Systems*, Vol.4, No.3, pp.165-184.

Ifinedo, P. & Nahar, N. (2006). Quality, impact and success of ERP systems: a study involving some firms in the Nordic-Baltic region. *Journal of Information Technology Impact*, Vol.6, pp.19-46.

Jensen, R. (1982). Adoption and diffusion of an innovation of uncertain profitability. *Journal of Economic Theory*, Vol.27, pp.182-193.

Jorgenson, D.W. & Stiroh, K.J. (1995). Computers and growth. *Economics Innovative New Technology*, Vol.3, No.4, pp.295–316.

Jorgenson, D.W. & Stiroh, K.J. (1999). Information Technology and Growth. *American Economic Review, Papers and Proceedings*, Vol.89, No.2, pp.109-115.

Katz, M.L. & Shapiro, C. (1986). Technology adoption in the presence of network externalities. *Journal of Political Economy*, August, pp.822-841.

Kiley, M.T. (1999). Computers and Growth With Costs of Adjustment: Will the Future Look Like the Past? *FEDS Working Paper No. 99-36*.

Kiley, M.T. (2001). Computers and growth with frictions: aggregate and disaggregate evidence. *Carnegie-Rochester Conference Series on Public Policy*, Vol.55, No.1, pp.171-215.

King, J.L. & Schrems, E.L. (1978). Cost-benefit analysis in information systems development and operation. *Computing Surveys*, Vol.10, No.1, pp.19-34.

Kraemer, K.L. & Dedrick, J. (1994). Payoffs from investment in information technology: Lessons from Asia-Pacific region. *World Development*, Vol.22, No.12, pp.19–21.

Kriebel, C.H., & Raviv, A. (1980). An Economics Approach to Modeling the Productivity of Computer Systems. *Management Science*, Vol.26, No.3, pp.297-311.

Lehr, W. & Lichtenberg, F.R. (1998). Computer use and productivity growth in the US federal government agencies, 1987-92. *Journal of Industrial Economics*, Vol.46, No.2, pp.257-279.

Lichtenberg, F.R. (1995). The output contributions of computer equipment and personnel: A firm level analysis. *Economics Innovations New Technology*, Vol.3, pp.201-217.

Litan, R.E. & Rivlin, A.M. (2001). Projecting the economic impact of the Internet. The *American Economic Review*, Vol.91, No.2, pp.313-317. Papers and Proceedings of the Hundred Thirteenth Annual Meeting of the American Economic Association (May).

Loveman, G.W. (1988). An assessment of the productivity impact on information technologies. MIT management in the 1990s. Working Paper #88-054.

Majumdar, S.K. & Venkataraman, S. (1998). Network effects and the adoption of new technology: Evidence from the US telecommunications industry. *Strategic Management Journal*, Vol.19, pp.1045-1062.

Markus, M.L. (1992). Critical mass contingencies for telecommunications consumers' In C. Antonelli (ed). The Economics of Information Networks. Elsevier, New York, pp.431-449.

Markus, M. L., Tanis, C. & Zmud, R. W. (2000). The enterprise systems experience-from adoption to success *Framing the Domains of IT Management: Projecting the Future Through the Past* (pp. 173-207): Pinnaflex Education Resources, Inc.

Matolcsy, Z.P., Booth, P., & Wieder, B. (2005). Economic benefits of enterprise resource planning systems: some empirical evidence. *Accounting & Finance*, Vol.45, No.3, pp. 439-456.

Morrison, C.J. & Siegel, D. (1997). External capital factors and increasing returns in US manufacturing. *The Review of Economics and Statistics*, Vol.79, No.4, pp.647-654.

Oliner, S.D. & Sichel, D.E. (1994). Computers and output growth revisited: how big is the puzzle? *Brooking Papers on Economic Activity*, Vol.2, pp.273-317.

Oliner, S.D. & Sichel, D.E. (2000). The resurgence of growth in the late 1990s: Is information technology the story?. *Journal of Economic Perspectives*, Vol.14, No.4, pp.3-22.

Poston, R., & Grabski, S. (2001). Financial impacts of enterprise resource planning implementations. *Information Systems*, Vol.2, pp.271-294.

Ragowsky, A. & Somers, T. M. (2002). Enterprise resource planning. *Journal of Management Information Systems*, Vol.19, No.1, pp.11-16.

Redmond, W.H. (1991). When Technologies Compete: The Role of Externalities in Nonlinear Market Response. *Journal of Product Innovation Management*, Vol.8, No.3, pp.170–183.

Roach, S.S. (1998). No productivity boom for workers. *Issues in Science and Technology*, Vol.14, No.4, pp.49-56.

Ross, J.W. & Vitale, M.R. (2000). The ERP Revolution: Surviving vs. Thriving. *Information Systems Frontiers*, Vol. 2, No. 2, pp. 233-241.

Shang, S. & Seddon, P.B. (2002). Assessing and managing the benefits of enterprise systems: the business manager's perspective. *Information Systems Journal*, Vol.12, No.4, pp.271-299.

Sharp, A.M., Register, C.A., and Grimes, P.W. (2006). Economics of Social Issues, 17th Ed. McGraw-Hill International Edition, New York: NY.

Siegel, D. (1997). The impact of computers on manufacturing productivity growth: a multiple-indicators multiple-causes appraoch. *The Review of Economics and Statistics*, Vol.79, No.1, pp.68-78.

Solow, R.M. (1987). We'd better watch out. *New York Times Book Review*, July 12. 36.

Spathis, C., & Ananiadis, J. (2005). Assessing the benefits of using an enterprise system in accounting information and management. *Journal of Enterprise Information Management*, Vol.18, No.2, pp.195-210.

Spathis, C., & Constantinides, S. (2003). The Usefulness of ERP Systems for Effective Management. *Industrial Management & Data Systems*, Vol. 103, No. 9, pp. 677-685.

Spathis, C., & Constantinides, S. (2004). Enterprise resource planning systems' impact on accounting processes. *Business Process Management Journal*, Vol.10, No.2, pp.234-247.

Stiroh, K.J. (1998). Computers, Productivity, and Input Substitution. *Economic Inquiry*, Vol. XXXVI, No.2, pp.175-191.

Su, Y, & Yang, C (2010). A structural equation model for analyzing the impact of ERP on SCM. *Expert Systems with Applications*, Vol.37, No.1, pp.456-469.

Varian, H., Litan, R.E., Elder, A., & Shutter, J. (2002). The net impact study: The projected economic benefits of the Internet in the United States, United Kingdom, France and Germany. http://www.netimpactstudy.com/NetImpact_Study_Report.pdf

Volkoff, O., Strong, D.M., & Elmes, M.B. (2005). Understanding enterprise systems-enabled integration. *European Journal of Information Systems*, Vol.14, No.2, pp.110-120.

Wagner, M., Phu, N.V., Azomahou, T., and Wehrmeyer, W. (2002). The relationship between the environmental and economic performance of firms: An empirical analysis of the European paper industry. Corporate Social Responsibility and Environmental Management, Vol. 9, pp.133-146.

Wieder, B., Booth, P., Matolcsy, Z.P., & Ossimitz, M. (2006). The impact of ERP systems on firm and business process performance. *Journal of Enterprise Information Management,* Vol.19, No.1, pp.13-29.

Wier, B., Hunton, J. & HassabElnaby, H. (2007). Enterprise resource planning systems and non-financial performance incentives: The joint impact on corporate performance. *International Journal of Accounting Information Systems,* Vol.8, No.3, pp.165-190.

Young, A.A. (1928). Increasing returns and economic progress. *Economuc Journal,* Vol.38, pp.527-542.

Zeng, A.Z. & Pathak, B.K. (2003). Achieving information integration in supply chain managmeent through B2B e-hub: concepts and analyses. *Industrial Manaement & Data Systems,* Vol.103, No.9, pp.657-665.

Zhu, Z., Scheuermmann, L. & Babineaux, B.J. Jr. (2004). Information network technology in the banking industry. *Industrial Management & Data Systems,* Vol.104, No.5, pp.409-417.

Part 2

Process Strategy and Process Analysis

4

Game Theoretic Analysis of Standby Systems

Kjell Hausken
Faculty of Social Sciences
University of Stavanger
Stavanger
Norway

1. Introduction

Systems have traditionally been analyzed assuming only one player, i.e. the defender maximizing system reliability facing exogenously fixed factors related to technology, nature, weather, culture, etc. After the September 11, 2001 attack realizations emerge that whereas one set of players work to ensure system reliability, another set of players oppose system reliability. This paper thus provides a game-theoretic analysis of a main system and a standby system. Examples of systems are power supply, telecommunications systems, water supply, roads, bridges, tunnels, political and economic institutions, businesses, schools, hospitals, recreational facilities, and various assets.

Many strategic considerations exist for a main system and a standby system. The players can choose their efforts in the present, or in the future dependent on the outcome of the strategic interactions in the present. For example, the defender of a water supply system will be intent on protecting it in the present since if it fails, the standby system has to take over in the future.

Compared with the literature the contribution of this paper is to provide an understanding of how the strategic interactions in the present are linked to the strategic interactions in the future for a main system and a standby system. This is done by analyzing a defender and an attacker in a two period game. The first period expresses the present and the second period expresses the future. The main system can fail dependent on the strategic interactions in the first period. This in turn impacts the strategic interactions in the second period. Furthermore, looking ahead to the second period before the game starts influences the players' strategies in the first period.

Systems where the future state of affairs depend on the present state of affairs are referred to as dependent systems (Ebeling 1997). Dependent systems have a long tradition of being analyzed using Markov analysis, which is unrelated to game theory.[1] A simple definition of

---

[1] First the number of system states is specified. Second the reliability is determined based on the system configuration. Third a rate diagram is designed where each node represents a state and each branch with an arrow specifies a transition rate (failure rate) expressed with a parameter. Fourth an equation is formulated for the probability of being in each state at time t+Δt which equals the probability of being in the state at time t, adding or subtracting the probabilities of moving into or out of the state from neighboring states when accounting for the transition rates. Fifth each equation is reformulated as a

a stochastic process with the Markov property is that the conditional probability distribution of future states of the process depends only upon the present state.[2] Markov analysis has proven highly successful applied to reliability analysis. This paper is concerned with two limitations of Markov analysis. First, enabling each of two players to choose strategies in each of two periods violates the Markov property since players are free to choose future strategies that are not conditioned on their present strategies. Generally, any theory involving intentional action (e.g. game theory) violates the Markov property. Second, this paper relaxes the constraint in Markov modeling where the transition rates between different states are kept constant through time.

This paper enables players to exert efforts to impact the system reliability as time progresses. That is, we analyze how players choose strategies through time to impact the reliability of dependent systems. We consider a dependent system consisting of a main system and a standby system. Both the main system and the standby system can be in two states which are to operate or fail. The dependent system is analyzed for general parameter values with backward induction as a two period game. We determine how two players make strategic decisions through time to impact the system reliability. Players allocate resources in the sense of substituting efforts across components and across time. Determining the nature of such substitutions is of substantial interest, see e.g. Enders and Sandler (2003), Hausken (2006), and Keohane and Zeckhauser (2003).

The paper answers questions such as whether players exert high efforts in the first period to position themselves for the second period, whether they are so weakened that they withdraw from the game, or whether they prefer the game to last one period or two periods. One player, the defender, maximizes the system reliability. The other player, the attacker, minimizes the system reliability. Both the main system's and the standby system's reliabilities depend on the relative levels of defense and attack and on the contest intensities. Each player's utility depends additively on the system reliability in the two periods, with a discount parameter varying between 0 and 1 for the second period. The unit costs of defense and attack, and the contest intensities, are different for the main system and the standby system, analogously to failure rates being different in Markov analysis dependent on the system state.

Hausken (2010) analyzes complex systems applying game theory. Hausken (2011) provides a game theoretic analysis of a two period dependent system of two components which can be fully operational, in two states of intermediate degradation, or fail.

For multi-state system reliability, see Lisnianski and Levitin (2003). For degraded systems see Ebeling's (1997:117ff) Markov analysis of a system which can be fully operational,

---

differential equation. The number of equations equals the number of states minus 1. The probability of being in the last state equals 1 minus the sum of the probabilities of being in the other states. Examples of systems analyzed with Markov analysis are load sharing systems, standby systems, degraded systems, and multistate systems (Ebeling 1997:108ff). For example, if one component fails in a load sharing system, the failure rates increase for the remaining components. A standby component may experience a low or zero failure rate in its standby state, and a higher failure rate when operational (which may or may not equal the failure rate of the originally operating component).

[2] Another definition is that of memorylessness where, conditional on the present state of the system, its future and past are independent. See e.g. Taylor and Karlin (1998) for further definitions.

degraded, ot failed. See Zio and Podofillini (2003) for Monte Carlo simulation of the effects of different system performance levels on the importance of multi-state components. Ramirez-Marquez and Coit (2005) use Monte-Carlo simulation to approximate multi-state two-terminal reliability. A next step is to incorporate strategic defenders and attackers into the analysis of multi-state and degraded systems.

In earlier research Levitin (2007) considers the optimal element separation and protection in a complex multi-state series-parallel system, and suggests an algorithm for determining the expected damage caused by a strategic attacker. Hausken and Levitin (2009) present a minmax optimization algorithm. The defender minimizes the maximum damage the attacker can inflict thereafter. The defender has multiple defense strategies which involve separation and protection of system elements. The attacker also has multiple attack strategies against different groups of system elements. A universal generating function technique is applied for evaluating the losses caused by system performance reduction. Levitin and Hausken (2009) introduce three defensive measures, i.e. providing redundancy, protecting genuine elements and deploying false elements and analyze the optimal resource distribution among these measures in parallel and k-out of-N systems. Levitin (2009) considers optimizing defense strategies for complex multi-state systems.

Azaiez and Bier (2007) consider the optimal resource allocation for security in reliability systems. Bier et al. (2005) analyze the protection of series and parallel systems with components of different values. They specify optimal defenses against intentional threats to system reliability, focusing on the tradeoff between investment cost and security. Bier et al. (2006) assume that a defender allocates defense to a collection of locations while an attacker chooses a location to attack. Hausken (2008) considers defense and attack for series and parallel reliability systems. Dighe et al. (2009) consider secrecy in defensive allocations as a strategy for achieving more cost-effective attacker deterrence.

Section 2 presents the model. Section 3 solves the model. Section 4 analyzes three special cases. Section 5 simulates the solution. Section 6 considers examples. Section 7 concludes.

2. The model

Consider a main system and a standby system. A defender and an attacker play a two period game. In period j, j=1,2, the defender exerts effort t_{Mj} at unit cost c_M to defend the main system, where t_{Mj} is the defender's free choice variable. Analogously, the attacker exerts effort T_{Mj} at unit cost C_M to attack the main system, where T_{Mj} is the attacker's free choice variable. If the main system is successfully defended in period 1 it continues to operate in period 2, while the standby system continues to be on standby. If the main system is successfully attacked in period 1 it does not operate in period 2, and instead the standby system operates in period 2. This means that the defender exerts effort t_{S2} at unit cost c_S to defend the standby system, and the attacker exerts effort T_{S2} at unit cost C_S to attack the standby system. Defense and attack are interpreted broadly. Defense means protecting against the attack, and maintaining and adjusting the system to prevent that it breaks down. Attacking means attacking the system, which may get aided by natural factors (technology, weather, temperature, humidity, etc.) to ensure that the system breaks down. We assume that the standby system does not operate in period 1 but is located in a secure place (e.g. underground bunker) from which it can be accessed if it is needed in period 2. We do not model efforts the defender and attacker may exert with respect to the standby system in period 1. In both periods both players make their strategic choices simultaneously and independently. Before the second period both players know the strategies chosen and the outcome of the first period. We formulate the reliability p_{Mj} of the main system in period j, and the reliability p_{S2} of the standby system in period 2, as contests between the defender and attacker. The most common functional form is the ratio form (Tullock 1980)

$$p_{Mj} = \frac{t_{Mj}^{m_M}}{t_{Mj}^{m_M} + T_{Mj}^{m_M}}, j = 1,2, \quad p_{S2} = \frac{t_{S2}^{m_S}}{t_{S2}^{m_S} + T_{S2}^{m_S}}, \tag{1}$$

where $\partial p_{Mj} / \partial t_{Mj} > 0$, $\partial p_{Mj} / \partial T_{Mj} < 0$, $\partial p_{S2} / \partial t_{S2} > 0$, $\partial p_{S2} / \partial T_{S2} < 0$, and $m_I \geq 0$, $I = M,S$ is a parameter for the contest intensities of the main system (M) and standby system (S). The reliabilities p_{Mj} and p_{S2} can also be interpreted as probabilities of system survival. Equation (1) is common in the rent seeking literature where the rent is an asset which corresponds to reliability in this paper. Conflict exists over reliability between the defender and the attacker, just as conflict exists over a rent between contending players. See Tullock (1980) for the use of m_I, Skaperdas (1996) for an axiomatization, Nitzan (1994) for a review, Hirshleifer (1995) for illustration, usefulness, and application, and Hausken (2005) for recent literature. At the limit, with infinitely much defensive effort, and finite offensive effort, system I is 100% reliable. The same result follows with finite defensive effort and zero offensive effort. At the other limit, with infinitely much offensive effort, and finite defensive effort, component i is 0% reliable. The same result follows with finite offensive effort and zero defensive effort. The sensitivity of p_{ij} to t_{ij} increases as m_I increases. When $m_I = 0$, the efforts t_{ij} and T_{ij} have equal impact on the reliability regardless of their size which gives 50% reliability, $p_{ij} = 1/2$. $0 < m_I < 1$ gives a disproportional advantage of exerting less effort than one's opponent. When $m_I = 1$, the efforts have proportional impact on the reliability. $m_I > 1$ gives a disproportional advantage of exerting more effort than one's opponent. This is often realistic in praxis, as evidenced by benefits from economies of scale. Finally, $m_I = \infty$ gives a step function where "winner-takes-all".

The main system can in period 1 be in the two states shown in Table 1, where v_M and V_M are the defender's and attacker's values of an operational main system given presence of a standby system.

State	Main system	Reliability	Defender value	Attacker value
1	operates	p_{M1}	$v_M - c_M t_{M1}$	$-C_M T_{M1}$
2	fails	$1 - p_{M1}$	$-c_M t_{M1}$	$V_M - C_M T_{M1}$

Table 1. Main system in two states in period 1.

We express the players' period 1 utilities as

$$u_1 = v_M p_{M1} - c_M t_{M1},$$
$$U_1 = V_M(1 - p_{M1}) - C_M T_{M1} \tag{2}$$

The players' period 1 strategic choices determine both their first period utilities and the system state as the start of period 2. Each time period can be short or long, e.g. one minute, one month, one shift, one season. If the main system fails in period 1, it cannot be repaired in time for the onset of period 2. This assumption is justified since repairing or replacing failed components can be complicated for economical and logistical reasons, and may require competence and time, which we assume is impossible both during the periods and in the transition from period 1 to period 2.[3] Hence the strategies the players choose for period 1 have to account for the combinations of possibilities in which the main system may operate or fail in the two periods, and the standby system may operate or fail in period 2. If the main system fails in period 1, since it cannot be repaired or replaced before the commencement of period 2, the players need to assess their defense and attack in period 2 to account for which of the two states (operation or failure) follows after period 1. In period 2 the players also make their strategic choices simultaneously and independently, knowing the outcome and choices in period 1.

If the main system operates after period 1 (state 1), then the unit costs of defense and attack remain unchanged and the players make strategic choices t_{M2} and T_{M2}. If the main system fails in period 1 (state 2), we assume that the unit costs of defense and attack for the standby system are c_S and C_S, and the contest intensity is m_S. The defender and attacker values of the standby system are v_S and V_S, where $v_S \leq v_M$ and $V_S \leq V_M$. The defender and attacker values after period 2 are shown in Table 2, where δ and Δ are time discount parameters.

State	Main system after period 1	Reliability after period 1	Defender value in period 2	Attacker value in period 2
1	operates	p_{M1}	$\delta(v_M p_{M2} - c_M t_{M2})$	$\Delta(V_M(1-p_{M2}) - C_M T_{M2})$
2	fails	$1 - p_{M1}$	$\delta(v_S p_{S2} - c_S t_{S2})$	$\Delta(V_S(1-p_{S2}) - C_S T_{S2})$

Table 2. Defender and attacker values after period 2.

We thus express the players' utilities over the two periods as

$$u = v_M p_{M1} - c_M t_{M1} + \delta p_{M1}(v_M p_{M2} - c_M t_{M2}) + \delta(1-p_{M1})(v_S p_{S2} - c_S t_{S2}),$$
$$U = V_M(1-p_{M1}) - C_M T_{M1} + \Delta p_{M1}(V_M(1-p_{M2}) - C_M T_{M2}) + \Delta(1-p_{M1})(V_S(1-p_{S2}) - C_S T_{S2}) \tag{3}$$

The third term on the right hand side in both utilities contains p_{M1} which is the probability that the main system survives period 1. The fourth and rightmost term on the right hand side in both utilities contains $1-p_{M1}$ which is the probability that the main system does not survive period 1. Fig. 1 shows the two-period game as an extensive form game tree.

3. Solving the model

The two players have two strategic choice variables t_{M1} and T_{M1} in period 1, and four strategic choice variables t_{M2}, T_{M2}, t_{S2}, and T_{S2} in period 2. We analyze pure strategy Nash equilibria. We solve the game with backward induction starting with period 2. Differentiating gives $\partial u / \partial t_{M2} = \partial u / \partial t_{S2} = 0$ and $\partial U / \partial T_{M2} = \partial U / \partial T_{S2} = 0$. Solving the four equations gives

---

[3] Future research may model the repair of the main system.

$$T_{12} = \frac{m_I \left(\frac{C_I/V_I}{c_I/v_I}\right)^{m_I} V_I}{C_I \left(1 + \left(\frac{C_I/V_I}{c_I/v_I}\right)^{m_I}\right)^2}, \quad t_{12} = \frac{C_I/V_I}{c_I/v_I} T_{12}, \quad p_{12} = \frac{1}{1 + \left(\frac{C_I/V_I}{c_I/v_I}\right)^{m_I}}, \quad I = M,S$$

$$u_{12} = v_I p_{12} - c_I t_{12} = \frac{1 + (1-m_I)\left(\frac{C_I/V_I}{c_I/v_I}\right)^{m_I}}{\left(1 + \left(\frac{C_I/V_I}{c_I/v_I}\right)^{m_I}\right)^2} v_I, \tag{4}$$

$$U_{12} = V_I(1 - p_{12}) - C_I T_{12} = \frac{1 - m_I + \left(\frac{C_I/V_I}{c_I/v_I}\right)^{m_I}}{\left(1 + \left(\frac{C_I/V_I}{c_I/v_I}\right)^{m_I}\right)^2} \left(\frac{C_I/V_I}{c_I/v_I}\right)^{m_I} V_I$$

where u_{12} and U_{12} are the period 2 utilities for system I, $I=M,S$. The second order conditions are satisfied when

$$\left(\frac{m_I - 1}{m_I + 1}\right)^{1/m_I} < \frac{C_I/V_I}{c_I/v_I} < \left(\frac{m_I + 1}{m_I - 1}\right)^{1/m_I}, I = M,S \tag{5}$$

Equation (5) is satisfied with an infinitely large range for the commonly used contest intensity $m_j=1$, and generally stretches from below to above $\dfrac{C_I/V_I}{c_I/v_I} = 1$.

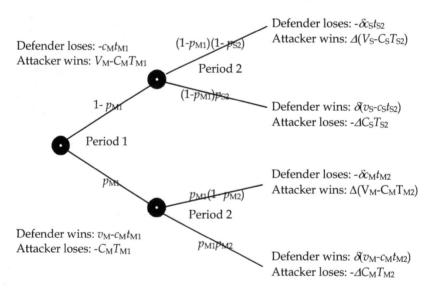

Fig. 1. The two-period extensive form game tree.

Equations (3) and (4) show that the second period strategic choice variables do not depend on the first period strategic choice variables, only on the parameters. This means that the two-period game gives the same result as a corresponding one-period game where the players choose their six strategies simultaneously and independently. This result follows since the players' strategic choices in period 2 are independent for the two states that are possible after period 1. For state 1 the strategic choice variables are t_{M2} and T_{M2}. For state 2 the strategic choice variables are t_{S2} and T_{S2}. Consequently the players do not need to know the outcome of period 1 in order to play period 2. But, the probabilities of the two states depend on how period 1 is played, so the players account for the outcome of period 2 for each of the two states when determining their strategies in period 1. Thus the expressions for t_{M2} and T_{M2} are valid for a one period main system as described in Table 1, and the expressions for t_{S2} and T_{S2} are valid for a one period system of a standby system.

To solve period 1 we rewrite (3) as

$$u = ap_{M1} + b - c_M t_{M1}, \quad U = A(1-p_{M1}) + B - C_M T_{M1},$$
$$a = v_M + \delta(v_M p_{M2} - c_M t_{M2} - v_S p_{S2} + c_S t_{S2}), \quad b = \delta(v_S p_{S2} - c_S t_{S2}), \tag{6}$$
$$A = V_M + \Delta(-V_M(1-p_{M2}) + C_M T_{M2} + V_S(1-p_{S2}) - C_S T_{S2}), \quad B = \Delta(V_M(1-p_{M2}) - C_M T_{M2})$$

where a,b,A,B are parameters determined by inserting (4) into (6). We interpret a and A as the defender's and attacker's values of an operational two period system consisting of a main system and a standby system. The parameters b and B provide direct values to the defender and attacker, respectively. Differentiating gives $\partial u / \partial t_{M1} = \partial U / \partial T_{M1} = 0$. Solving the two equations gives

$$T_{M1} = \frac{m_M \left(\dfrac{C_M/A}{c_M/a}\right)^{m_M} A}{C_M \left(1 + \left(\dfrac{C_M/A}{c_M/a}\right)^{m_M}\right)^2}, \quad t_{M1} = \frac{C_M/A}{c_M/a} T_{M1}, \quad p_{M1} = \frac{1}{1 + \left(\dfrac{C_M/A}{c_M/a}\right)^{m_M}},$$

$$u_1 = \frac{v_M + (v_M - am_M)\left(\dfrac{C_M/A}{c_M/a}\right)^{m_M}}{\left(1 + \left(\dfrac{C_M/A}{c_M/a}\right)^{m_M}\right)^2}, \quad U_1 = \frac{V_M - Am_M + V_M \left(\dfrac{C_M/A}{c_M/a}\right)^{m_M}}{\left(1 + \left(\dfrac{C_M/A}{c_M/a}\right)^{m_M}\right)^2} \left(\dfrac{C_M/A}{c_M/a}\right)^{m_M}, \tag{7}$$

$$u = \frac{1 + (1 - m_M)\left(\dfrac{C_M/A}{c_M/a}\right)^{m_M}}{\left(1 + \left(\dfrac{C_M/A}{c_M/a}\right)^{m_M}\right)^2} a + b, \quad U = \frac{1 - m_M + \left(\dfrac{C_M/A}{c_M/a}\right)^{m_M}}{\left(1 + \left(\dfrac{C_M/A}{c_M/a}\right)^{m_M}\right)^2} \left(\dfrac{C_M/A}{c_M/a}\right)^{m_M} A + B$$

where

$$a = v_M + \delta \left(\frac{1+(1-m_M)\left(\frac{C_M/V_M}{c_M/v_M}\right)^{m_M}}{\left(1+\left(\frac{C_M/V_M}{c_M/v_M}\right)^{m_M}\right)^2} v_M - \frac{1+(1-m_S)\left(\frac{C_S/V_S}{c_S/v_S}\right)^{m_S}}{\left(1+\left(\frac{C_S/V_S}{c_S/v_S}\right)^{m_S}\right)^2} v_S \right),$$

$$A = V_M + \Delta \left(\frac{1-m_S+\left(\frac{C_S/V_S}{c_S/v_S}\right)^{m_S}}{\left(1+\left(\frac{C_S/V_S}{c_S/v_S}\right)^{m_S}\right)^2} \left(\frac{C_S/V_S}{c_S/v_S}\right)^{m_S} V_S - \frac{1-m_M+\left(\frac{C_M/V_M}{c_M/v_M}\right)^{m_M}}{\left(1+\left(\frac{C_M/V_M}{c_M/v_M}\right)^{m_M}\right)^2} \left(\frac{C_M/V_M}{c_M/v_M}\right)^{m_M} V_M \right), \quad (8)$$

$$b = \delta \frac{1+(1-m_S)\left(\frac{C_S/V_S}{c_S/v_S}\right)^{m_S}}{\left(1+\left(\frac{C_S/V_S}{c_S/v_S}\right)^{m_S}\right)^2} v_S, \quad B = \Delta \frac{1-m_M+\left(\frac{C_M/V_M}{c_M/v_M}\right)^{m_M}}{\left(1+\left(\frac{C_M/V_M}{c_M/v_M}\right)^{m_M}\right)^2} \left(\frac{C_M/V_M}{c_M/v_M}\right)^{m_M} V_M$$

Inserting $\delta=\Delta=0$ into (8) gives $a=v_M$, $A=V_M$, $b=B=0$, $T_{M1}=T_{M2}$, and $t_{M1}=t_{M2}$, confirming that period 1 operates as the last period 2 with maximum discounting. Equation (8) implies $\lim_{m_M \to \infty} a = 0$ (since a cannot be negative) and $\lim_{m_M \to \infty} A = \infty$ and thus high contest intensity for the main system causes the defender and attacker to perceive the two period system as valueless and very valuable, respectively. Conversely, (8) also implies $\lim_{m_S \to \infty} a = \infty$ (since a cannot be negative) and $\lim_{m_S \to \infty} A = 0$ (since A cannot be negative) and thus high contest intensity for the standby system causes the defender and attacker to perceive the two period system as very valuable and valueless, respectively.

The second order conditions are satisfied when

$$\left(\frac{m_M-1}{m_M+1}\right)^{1/m_M} < \frac{C_M/A}{c_M/a} < \left(\frac{m_M+1}{m_M-1}\right)^{1/m_M} \tag{9}$$

The boundary solutions are as follows. The interior solution above is valid when $u \geq 0$ and $U \geq 0$ in (7). When $u < 0$ or $U < 0$, no pure-strategy Nash equilibrium exists. Analyzing mixed strategy equilibria is beyond the scope of this paper. Equation (7) and the next sections show that $u < 0$ and $U < 0$ are possible when the contest intensity of the main system m_M is large which induces large costly efforts. The case $u < 0$ is calamitous for the defender since it cannot earn positive utility. We assume that the defender withdraws in this case, exerting zero effort and earning zero utility, while the attacker exerts negligible effort earning utility $V_M + \Delta V_S$ since all reliabilities are zero. One may reason that if the defender knows that the attacker exerts negligible effort, the defender can exert positive effort and earn positive utility. However, if the attacker knows that, it can exert positive effort and earn positive utility. In the absence of a pure-strategy Nash equilibrium, the assumption of withdrawal is plausible. Analogously, for the case $U<0$, we assume that the attacker withdraws exerting zero effort and earning zero utility, while the defender exerts negligible effort earning utility $(1+\delta)v_M$ since all reliabilities are one.

4. Analyzing three special cases

Let us consider three special benchmark cases with straightforward interpretations. The first is the egalitarian case $m_M=m_S=0$ causing zero efforts and thus 50% probability of failure for the main system in period 1, and for the main system or standby system in period 2. This case illustrates how the players' utilities depend on the main system and standby system and the weights assigned to period 2 expressed with δ and Δ. Cases 2 and 3 assume $\dfrac{C_I / V_I}{c_I / v_I}$ =1, which occurs e.g. when the players have equal unit costs $C_I=c_i$ and evaluations $V_I=v_i$, and equal contest intensities for the main system and the standby system, $m_M=m_S=m$. For case 2 we show how this impacts the values a,A,b,B of the two period system dependent on the discount parameters δ and Δ. For case 3 we furthermore show the impact of no discounting $\delta=\Delta=1$.

First, inserting $m_M=m_S=0$ into (4),(7),(8) gives

$$T_{M1} = t_{M1} = T_{I2} = t_{I2} = 0, \quad p_{M1} = p_{I2} = \frac{1}{2}, \quad I = M,S,$$

$$u_1 = \frac{v_M}{2}, U_1 = \frac{V_M}{2}, u_{I2} = \frac{v_I}{2}, U_{I2} = \frac{V_I}{2}, \quad u = \frac{v_M}{2} + \frac{\delta}{4}(v_M + v_S), \quad U = \frac{V_M}{2} + \frac{\Delta}{4}(V_S + V_M) \tag{10}$$

The utilities are not affected by efforts in egalitarian contests, so the players choose zero efforts which cause 50% reliability for the main system in period 1. The main system has 50% probability of surviving into period 2, and thus 25% reliability at the end of period 2. The standby system has 50% probability of being implemented in period 2, and thus 25% reliability at the end of period 2. The utilities are positive in both periods.

Second, inserting $\dfrac{C_I / V_I}{c_I / v_I}$ =1 and $m_M=m_S=m$ into (8) gives

$$a = v_M + \delta\frac{2-m}{4}(v_M - v_S), \quad A = V_M + \Delta\frac{2-m}{4}(V_S - V_M), \quad b = \delta\frac{2-m}{4}v_S, \quad B = \Delta\frac{2-m}{4}V_M \tag{11}$$

giving rise to three observations. 1. The defender gets increased value of the two period system and the attacker gets decreased value of the two period system with low contest intensity $m<2$, since $v_S \le v_M$ and $V_S \le V_M$. High contest intensity $m>2$ causes the reverse result and is costly for the defender. 2. Contest intensity $m=2$ gives $a=v_M$, $A=V_M$, $b=B=0$, $T_{M1}=T_{M2}$, and $t_{M1}=t_{M2}$. 3. Equal values $v_S=v_M$ and $V_S=V_M$ give $a=v_M$, $A=V_M$, $T_{M1}=T_{M2}$, and $t_{M1}=t_{M2}$.

Third, inserting $\dfrac{C_I / V_I}{c_I / v_I}$ =1 and $m_M=m_S=m=\delta=\Delta=1$ into (4),(7),(8) gives

$$T_{M1} = \frac{\dfrac{C_M/A}{c_M/a}A}{C_M\left(1+\dfrac{C_M/A}{c_M/a}\right)^2}, \quad t_{M1} = \frac{C_M/A}{c_M/a}T_{M1}, \quad p_{M1} = \frac{1}{1+\dfrac{C_M/A}{c_M/a}},$$

$$T_{I2} = t_{I2} = \frac{V_I}{4C_I}, \quad p_{I2} = \frac{1}{2}, \quad I = M,S,$$

$$u_1 = \frac{v_M + (v_M - a)\dfrac{C_M/A}{c_M/a}}{\left(1+\dfrac{C_M/A}{c_M/a}\right)^2}, \quad U_1 = \frac{V_M - A + V_M\dfrac{C_M/A}{c_M/a}}{\left(1+\dfrac{C_M/A}{c_M/a}\right)^2}\frac{C_M/A}{c_M/a},$$

$$u_{I2} = \frac{v_I}{\left(1+\dfrac{C_I/V_I}{c_I/v_I}\right)^2}, \quad U_{I2} = \frac{\left(\dfrac{C_I/V_I}{c_I/v_I}\right)^2}{\left(1+\dfrac{C_I/V_I}{c_I/v_I}\right)^2}V_I,$$

$$u = \frac{a}{\left(1+\dfrac{C_M/A}{c_M/a}\right)^2}+b, \quad U = \frac{\left(\dfrac{C_M/A}{c_M/a}\right)^2 A}{\left(1+\dfrac{C_M/A}{c_M/a}\right)^2}+B, \tag{12}$$

$$a = \frac{5v_M - v_S}{4}, \quad A = \frac{3V_M + V_S}{4}, \quad b = \frac{v_S}{4}, \quad B = \frac{V_M}{4}$$

5. Simulating the solution

Figs. 2-5 plot the six efforts $t_{M1}, T_{M1}, t_{M2}, T_{M2}, t_{S2}, T_{S2}$ and two utilities u and U as functions of one parameter relative to the baseline $c_M=C_M=c_S=C_S=v_M=V_M=v_S=V_S=m_M=m_S=\delta=\Delta=1$. The titles on the vertical axis are as specified in the legend box. Fig. 1 panel 1 plots as functions of the standby system values $v_S=V_S$ varying between 0 and 1. When the standby system has its maximum value $v_S=V_S=v_M=V_M=1$, the six efforts equal 0.25 and the utilities are 0.5, as also seen from (12) where $a=A=1$ and $b=B=0.25$. As $v_S=V_S$ decrease below 1, the efforts $t_{S2}=T_{S2}$ decrease linearly toward zero, while $t_{M2}=T_{M2}$ remain constant at 0.25. However, the defender compensates for the decreased value of the standby system by defending the main system more thoroughly in period 1, and thus t_{M1} increases when $v_S=V_S$ decrease. The attacker responds to this by decreasing T_{M1} when $v_S=V_S$ decrease, and thus the attacker's utility U also decreases when $v_S=V_S$ decrease. The defender's utility u is almost constant (slightly U shaped) since the defender compensates for the decreasing value of the standby system by defending the main system more thoroughly in period 1. Fig. 1 panel 2 plots as functions of the main system values $v_M=V_M$ increasing upwards from $v_S=V_S=1$. Now the standby system efforts $t_{S2}=T_{S2}$ remain constant at 0.25, and all the other variables increase in $v_M=V_M$. Increasing $v_M=V_M$ relative to the fixed $v_S=V_S=1$ induces the defender to defend the more valuable main system more thoroughly in period 1, and thus t_{M1} increases more than T_{M1}, and the defender's utility u increases more than the attacker's utility U, though $t_{M2}=T_{M2}$ for the main system in period 2 increase equivalently.

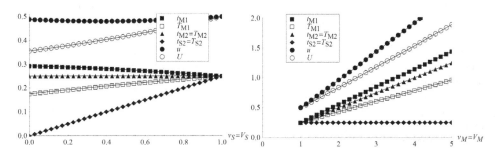

Fig. 2. Efforts $t_{M1}, T_{M1}, t_{M2}, T_{M2}, t_{S2}, T_{S2}$ and utilities u and U as functions of $v_S = V_S$ and $v_M = V_M$.

Fig. 3 panel 1 plots as functions of the defender's unit costs $c_M = c_S$ changing equally for both systems. The three defender efforts $t_{M1} = t_{M2} = t_{S2}$ decrease convexly and equivalently as the defense becomes more costly. The three attacker efforts $T_{M1} = T_{M2} = T_{S2}$ are inverse U shaped. When $c_M = c_S$ is low, the inferior attacker provides modest efforts against the defender cheaply producing a substantial defense. When $c_M = c_S$ is high, the defender efforts are low and the attacker does not need to attack substantially. Hence the attacker efforts are largest for intermediate $c_M = c_S$. The defender utility decreases, and the attacker utility increases, in $c_M = c_S$. Fig. 3 panel 2 plots as functions of the defender's unit cost c_M for the main system, keeping $c_S = 1$ for the standby system. The results are similar but $t_{S2} = T_{S2}$ remain constant at 0.25. When $c_M < 1$, both the defender's and the attacker's efforts for the main system are larger in period 1 than in period 2, and conversely when $c_M > 1$.

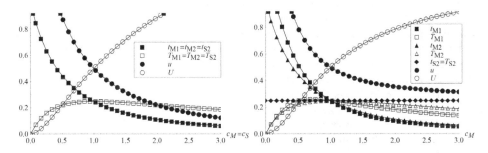

Fig. 3. Efforts $t_{M1}, T_{M1}, t_{M2}, T_{M2}, t_{S2}, T_{S2}$ and utilities u and U as functions of $c_M = c_S$ and c_M.

Fig. 4 panel 1 plots as functions of equivalent contest intensities $m_M = m_S$ for both systems. High contest intensities induce higher efforts which increase linearly in $m_M = m_S$. The higher efforts are costly causing the utilities to decrease linearly reaching zero when $m_M = m_S = 2$. Fig. 4 panel 2 plots as functions of the contest intensity m_M for the main system, keeping $m_S = 1$. Thus $t_{S2} = T_{S2}$ remain constant at 0.25. When $m_M < 1$, the results are similar but the comparatively higher $m_S = 1$ causes lower utilities in panel 2. For the main system in period 2 the efforts increase linearly in m_M as seen from (4) inserting $c_M = C_M = v_M = V_M = 1$. However, for the main system in period 1 the efforts do not increase linearly since A and a, instead of v_M and V_M, operate in the efforts t_{M1} and T_{M1} in (7). As m_M increases above 1, t_{M1} and T_{M1} increase in a decreasing manner, reaching maxima and thereafter decreasing towards zero.

The defender accepts its maximum for t_{M1} being lower than the attacker's maximum for T_{M1}. This gives low reliability p_{M1} for the main system in period 1 as m_M increases above 1. The high contest intensity for the main system makes it too costly to defend and attack compared with defending and attacking the standby system in period 2. The defender's utility decreases below 0.25 as m_M increases above 1.61, reaches a minimum 0.14 for $m_M=2.22$ as the defender realizes that the main system is too costly to defend, and increases asymptotically towards 0.25. Recall from (3) that if the defender does not defend the main system in period 1, it is guaranteed to fail, $p_{M1}=0$, causing utility 0.25 generated by the standby system since $p_{S2}=0.5$ with the given parameter values. Hence the attacker's utility eventually increases towards 1.25.

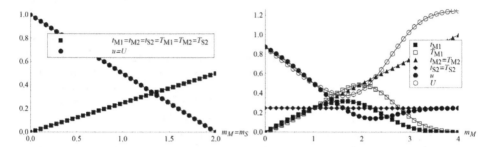

Fig. 4. Efforts $t_{M1},T_{M1},t_{M2},T_{M2},t_{S2},T_{S2}$ and utilities u and U as functions of $m_M=m_S$ and m_M.

Fig. 5 panel 1 plots as functions of the discount parameters $\delta=\Delta$. All efforts are constant at 0.25. The utilities increase from 0.25 when period 2 is discounted ($\delta=\Delta=0$) and reliability is $p_{M1}=0.5$ for period 1, to 0.5 when both periods have equal weight ($\delta=\Delta=1$) and all reliabilities are $p_{M1}=p_{M2}=p_{S2}=0.5$. Fig. 5 panel 2 also plots as functions of $\delta=\Delta$, but decreases the values of the standby system to $v_S=V_S=0.5$. This decreases the efforts for the standby system to $t_{S2}=T_{S2}=0.125$. The utilities remain at 0.25 when $\delta=\Delta=0$ and period 2 is irrelevant. As $\delta=\Delta$ increase, the defender compensates for the less valuable standby system by increasing t_{M1} from 0.25 to 0.28 when $\delta=\Delta=1$, which increases the reliability of the main system in period 1. The attacker decreases T_{M1} from 0.25 to 0.22 when $\delta=\Delta=1$, while $t_{M2}=T_{M2}$ remain constant at 0.25. The defender's utility thus increases more than the attacker's utility, but both utilities increase less than in panel 1 where the standby system is more valuable at $v_S=V_S=1$.

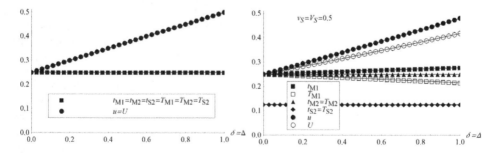

Fig. 5. Efforts $t_{M1},T_{M1},t_{M2},T_{M2},t_{S2},T_{S2}$ and utilities u and U as functions of $\delta=\Delta$.

6. Examples

There are no limits to the kinds of standby systems that can be envisioned. In fact, any system produced to deliver some function, can be supplemented with a standby system to deliver the same function in the event that the main system breaks down. Examples of systems are power supply, telecommunications systems, water supply, roads, bridges, tunnels, political and economic institutions, businesses, schools, hospitals, recreational facilities, and various assets. One example of a standby system is a standby generator which is a back-up electrical system that operates automatically (Hickey 2002). Within seconds of a utility outage, an automatic transfer switch senses the power loss, directs the standby generator to start, and transfers the electrical load to the standby generator. The standby generator thereafter supplies power to the circuits. To ensure a proper response to a power outage, a standby generator runs weekly self-tests. Most units run on diesel, natural gas or liquid propane gas. Automatic standby generators may be required by building codes for critical safety systems. Examples are building elevators, fire protection systems, standby lighting, or medical and life support equipment. Residential standby generators are common, providing backup electrical power to security systems, household appliances such as refrigerators, stoves, and hot water heaters, and HVAC systems. To determine the quality, design, and maintenance regime for the main system and standby system in each particular example, the analysis in this paper can be used.

7. Conclusion

We consider two players choosing strategies through time to impact the reliability of a dependent system which consists of a main system and a standby system. Each system can be in two states, i.e. it can operate or fail. If the main system operates successfully through period 1, it continues to operate into period 2 and the standby system remains in standby. If the main system fails in period 1, the standby system is implemented in period 2.

Each system is protected by a defender which maximizes its reliability subtracting the defense costs, and attacked by an attacker which maximizes its unreliability subtracting the attack costs. Each system's reliability depends on the relative levels of defense and attack and on the contest intensity. Each player's utility depends additively on the system reliability in two time periods, with a time discount parameter for the second period. The unit costs of effort and the contest intensities are different for the two players and the two systems. The two period game is analyzed with backward induction.

In period 1 the defender chooses a defense effort and the attacker chooses an attack effort for the main system. In period 2 the defender chooses one defense effort for the main system and one defense effort for the standby system, not knowing before the game starts whether the main system or the standby system is the system to be defended in period 2. Analogously in period 2, the attacker chooses one attack effort for the main system and one attack system for the standby system. Hence six strategic decisions are made by the two players.

The players assign different values to the main system and the standby system. We present analytical solutions and simulations to illustrate the players' efforts in the two periods and the utilities dependent on parametric changes.

Each time period can be short or long, e.g. minutes, days, months, as determined by the nature of the system and how failures occur dependent on the players' efforts. Since the main system fails in period 1 with positive probability, period 2 starts with the standby system with this positive probability.

Before period 1 each player assesses its effort for period 1 knowing that this effort impacts the probability that the main system survives into period 2. Exerting low defense effort in period 1 increases the probability that the standby system is implemented in period 2. We show that as the value of the standby system decreases below that of the main system, the defender increases its defense of the main system in period 1. Increasing the value of the main system increases the defender's utility more than the attacker's utility since the defender defends the main system more thoroughly in period 1. High defense effort in period 1 is an investment into the future for the defender.

As the defender's unit defense costs increases, its efforts and utilities decrease while the attacker's efforts are inverse U shaped and its utility increases. This follows since low unit defense costs make it not worthwile for the inferior attacker to attack, while high unit defense costs make it unnecessary for the superior attacker to attack substantially.

Increasing contest intensities for both systems causes all efforts to increase driving utilities downwards eventually reaching zero. Increasing the contest intensity only for the main system causes both efforts for the main system in period 1 to be inverse U shaped but taller for the attacker. This increases the probability that the main system fails in period 1. This benefits the attacker and does not benefit the defender which resorts to defending the standby system in period 2. Increasing discount parameters, making period 2 more valuable, benefit both players.

Two limitations of Markov analysis have been illustrated in this paper. First, we have enabled players to choose efforts strategically, which violates the Markov property. Second, we have relaxed the constraint in Markov modeling where the transition rates between different states are kept constant through time. The parameter values for the standby system in period 2 may differ from the parameter values for the main system in period 1. Future research may model in a multi-period game multiple states of operation for the main system and the standby system, and repair of the main system.

8. Notation

t_{Mj}	defender's effort to protect main system in period j, j=1,2
T_{Mj}	attacker's effort to attack main system in period j, j=1,2
t_{S2}	defender's effort to protect standby system in period 2
T_{S2}	attacker's effort to attack standby system in period 2
p_{Mj}	reliability of main system in period j
p_{S2}	reliability of standby system in period 2
c_M	defender's unit cost of effort for main system
C_M	attacker's unit cost of effort for main system
c_S	defender's unit cost of effort for standby system
C_S	attacker's unit cost of effort for standby system
v_M	defender's value of operational main system given presence of a standby system
V_M	attacker's value of operational main system given presence of a standby system

v_S	defender's value of standby system
V_S	attacker's value of standby system
m_M	attacker-defender contest intensity for main system
m_S	attacker-defender contest intensity for standby system
δ	defender's time discount parameter for period 2
Δ	attacker's time discount parameter for period 2
a	defender's value of an operational two period system
A	attacker's value of an operational two period system
b	additional value to defender of two period system
B	additional value to attacker of two period system
u	defender's utility
U	attacker's utility

9. References

Azaiez, N., Bier, V.M. (2007). Optimal Resource Allocation for Security in Reliability Systems. European Journal of Operational Research 181, 2, 773-786.

Bier, V.M., Nagaraj, A., Abhichandani, V. (2005). Protection of Simple Series and Parallel Systems with Components of Different Values. Reliability Engineering and System Safety 87, 315-323.

Bier, V.M., Oliveros, S., Samuelson, L. (2006). Choosing What to Protect: Strategic Defense Allocation Against an Unknown Attacker. Journal of Public Economic Theory 9, 4, 563-587.

Dighe, N., Zhuang J., and Bier V.M. (2009). Secrecy in defensive allocations as a strategy for achieving more cost-effective attacker deterrence, International Journal of Performability Engineering, special issue on System Survivability and Defense against External Impacts, 5, 1, 31-43.

Ebeling, C. (1997). An introduction to Reliability and Maintainability Engineering, McGraw-Hill, New York.

Enders, W., Sandler, T. (2003). What do we know about the substitution effect in transnational terrorism?. in A. Silke and G. Ilardi (eds) Researching Terrorism: Trends, Achievements, Failures (Frank Cass, Ilfords, UK), http://www-rcf.usc.edu/~tsandler/substitution2ms.pdf

Hausken, K., (2005) Production and conflict models versus rent seeking models. Public Choice 123, 1, 59-93.

Hausken, K. (2006), Income, Interdependence, and Substitution Effects Affecting Incentives for Security Investment, Journal of Accounting and Public Policy 25, 6, 629-665.

Hausken, K. (2007), Stubbornness, Power, and Equilibrium Selection in Repeated Games with Multiple Equilibria, Theory and Decision 62, 2, 135-160.

Hausken, K. (2008), Strategic Defense and Attack for Series and Parallel Reliability Systems, European Journal of Operational Research 186, 2, 856-881.

Hausken, K. (2010), Defense and Attack of Complex and Dependent System, Reliability Engineering & System Safety 95, 1, 29-42.

Hausken, K. (2011), Game Theoretic Analysis of Two Period Dependent Degraded Multistate Reliability Systems, International Game Theory Review Forthcoming.

Hausken, K. (2011), Protecting Complex Infrastructures Against Multiple Strategic Attackers, International Journal of Systems Science 42, 1, 11-29.

Hausken, K. and Levitin, G. (2009), Minmax defense strategy for complex multi-state systems, Reliability Engineering & System Safety 94, 2, 577-587.

Hickey, R.B. (2002). Electrical Construction Databook, McGraw Hill, New York, Chapter 14.

Hirshleifer, J. (1995). Anarchy and Its Breakdown. Journal of Political Economy 103, 1, 26-52.

Keohane N., Zeckhauser R.J. (2003). The ecology of terror defense. Journal of Risk and Uncertainty 26, 201-229.

Levitin, G., (2007). Optimal Defense Strategy Against Intentional Attacks. IEEE Transactions on Reliability 56, 1, 148-156.

Levitin, G. (2009). Optimizing Defense Strategies for Complex Multi-State Systems. In: Bier VM and Azaiez MN (eds.), Game Theoretic Risk Analysis of Security Threats, Springer, New York, 33-64.

Levitin, G. and Hausken, K. (2009), Redundancy vs. Protection vs. False Targets for Systems under Attack, IEEE Transactions on Reliability 58, 1, 58-68.

Lisnianski, A. and Levitin, G. (2003), Multi-state system reliability. Assessment, optimization and applications, World Scientific, New Jersey.

Nitzan, S. (1994), Modelling Rent-Seeking Contests, European Journal of Political Economy 10, 1, 41-60.

Ramirez-Marquez, J.E. and Coit, D.W., (2005). A Monte-Carlo simulation approach for approximating multi-state two-terminal reliability, Reliability Engineering & System Safety 87, 2, 253-264.

Skaperdas, S., (1996). Contest success functions. Economic Theory 7, 2, 283-290.

Taylor H.M., Karlin S., (1998), An Introduction To Stochastic Modeling, Third Edition, Academic Press, New York.

Tullock, G. (1980). Efficient Rent-Seeking. In Buchanan, J.M., Tollison, R.D., and Tullock, G., Toward a Theory of the Rent-Seeking Society, Texas A. & M. University Press, College Station, 97-112.

Zio, E. and Podofillini, L. (2003), Monte Carlo simulation analysis of the effects of different system performance levels on the importance of multi-state components, Reliability Engineering & System Safety 82, 1, 63-73.

5

A Value Structured Approach to Conflicts in Environmental Management

Fred Wenstøp

BI Norwegian Business School
Norway

1. Introduction

The paper aims to suggest a way for OR practitioners to approach value conflicts in environmental management. The approach is claimed to be practical and at the same time theoretically well founded on three pillars: ethics, neuro-economics, and decision sciences.

The aim springs out of the concern that a rational approach to environmental management problems too often is hampered or even perverted by strong emotions elicited by value conflicts among stakeholders. Thus, considerable resources are frequently squandered on ill founded projects that may have detrimental effects. This could be mitigated if the OR analyst has a deeper understanding of ethical reasons for choice, as well as knowledge of practical methods to deal with values.

2. OR relevance

Environmental management problems involve facts as well as values. It is not only a question of being able to predict the outcomes of different actions, but we also need to decide whether the outcomes are good or bad. OR embraces an arsenal of tools to predict or optimize environmental consequences of human action. This includes System Dynamics, which highlights causal relations and dynamic effects; with *The Limits to Growth* (Meadows et al., 1972) as the most celebrated application. OR also embraces the field of Multi Criteria Decision Analysis (MCDA) which is designed to deal with values. Thus, OR appears to be well equipped to handle environmental problems. This is also evident from the number of papers being published in OR outlets on environmental issues. A simple count (May 2011) of papers with the word "environmental" in the title, abstract or among the key-words runs to 536 in *Omega* and 1608 in the *European Journal of Operational Research*. The tallies are respectively 96 and 42 for "sustainability".

While MCDA has methods to weight decision criteria, it has less to offer when ethical issues transcend the mere comparison of values (Wenstøp, 2005). This typically happens in environmental management cases where conflicts of rights and sense of duty often preclude discussion of consequences. This paper addresses that problem.

3. OR literature on value conflicts

OR has not only a hard mathematical core, but also a rich tradition that emphasizes soft methods for structuring problems and to facilitate stakeholder involvement. Jonathan Rosenhead, with his Problem Structuring Methods, is one notable champion (Rosenhead, 2005). The current paper takes a more analytic approach than this, however. It has an emphasis on values and ethics, which can be traced as another OR tradition (Brans and Gallo, 2007), (Wenstøp, 2010), and within this tradition there are several publications on environmental management. One of the first contributors was Kenneth Boulding (1966) who coined the term "spaceship earth", signaling an early warning that energy, material, and environmental amenities are limited, and therefore require careful husbandry. More recently, Rauschmayer (2001) reflects on the normative foundation of MCDA and argue that decision criteria have to reflect not only the interests but possibly all values stemming from normative arguments of the decision-maker. This is especially true in environmental management where "the integration of values will result in changes of the MCA understanding, criteria building, and aggregation method, and will not be possible without analytical capacities of the decision analyst in ethics". Brans (2004) promotes Multi Criteria Decision Analysis as a suitable OR tool to take the interests of the stakeholders and Nature into account and calls for a multifaceted concept of ethics, consisting of Respect, Multi Criteria Management and Happiness. Kunsch (2009) discusses OR techniques to model decision-making problems with ethical dimensions, such as sustainability issues in the triangle of society, economy and environment, and Brans and Kunsch (2010) propose practical OR methods and tools for dealing with sustainability issues.

Le Menestrel and Van Wassenhove (2004), (2009) discuss the important issue of how to deal with the tension between the scientific legitimacy of OR models – where ethics is kept outside the models, and the integration of ethics within the models. The current paper is a voice in that debate. It takes Wenstøp and Koppang's (2009) view on OR and value conflicts as the point of departure and concludes that OR ought to handle decision problems involving value conflicts in environmental management by separating values according to ethical category.

4. Outline

When value conflicts arise in environmental management – which they often do – emotion laden arguments with an ethical undertone are notorious. I start the paper by elucidating this by narrating a dialogue regarding invasive aliens between a 'Socratic' journalist and a state employed environmental manager. I then propose to classify the arguments according to the classical ethical categories of virtue, duty and consequence. This makes it possible to set the conflict in a theoretical perspective by describing a consilience (Wilson, 1998) among the three ethical categories, three classes of values, and three kinds of emotions, and use that system to propose how the arguments can be organized and prepared for an OR approach to the problem. I finally recommend how OR ought to approach value laden decision problems in environmental management.

5. Case: The alien raccoon dog

The raccoon dog (*Nyctereutes procyonides*) is a small animal with short ears and a furry body. It enjoys high prestige in Japanese and Korean folklore where it is known as *Tanuki*, a merry and mischievous rascal, master of disguise, but a bit gullible. It looks somewhat like a racoon, but belongs to the same family as the dog. It was introduced to Russia from Korea around 1930 because of its fur, where after it migrated westward from Russia and reached Finnmark in northern Norway in 1983, where it has been observed six times during 2010. In Norway, the raccoon dog is officially considered alien, invasive and possibly detrimental to other species. It is considered a potential carrier of tapeworm and rabies. The Directorate for Nature Management (DNM) – one of five governmental agencies under the Ministry of Environment – developed an action plan (2008) to prevent its invasion of Norway. It has the form of a 17 page document with a front picture of an aggressive raccoon dog attacking a Norwegian magpie. It is presented as opportunistic, alien and harmful, a carrier of tapeworm and rabies, an alien species that must be exterminated.

The following is a narrative of an interview published in a Norwegian newspaper (Q) with an advisor in DNM (A) (Sætre, 2010). Since one important reason for exterminating the raccoon dog is that it may carry tapeworm this was a natural start of the dialogue:

Q: Dogs and foxes carry tapeworm as well? A: Right, but raccoon dogs wander more.

Q: Wolves carry rabies too? A: Yes, but raccoon dogs may spread rabies faster.

Q: The potato is also alien? A: All cultural plants are alien. We wanted them for food. Our civilization depends on this. But we think differently now...

Q: How many raccoon dogs are there in Norway? A: Two, for certain, but there may be more. And if we allow them to breed, they will soon threaten our Norwegian animals, who have lived in peace and harmony. Sitting ducks are especially vulnerable.

Q: Why is it more important for us to have ducks than raccoon dogs? A: Hunting; it is a traditional pastime to hunt ducks. So this is a value based choice.

Q: So what you mean is that our values decide for ducks and against raccoon dogs? A: Yes, and that is quite legitimate. Hunting traditions you know. The experience of having ducks around...

Q: So we might turn this around then and argue that the raccoon dog is valuable? A: Yes, you are free to do that, but the Norwegian policy is to prevent it from establishing itself. If it does, we may not even pick blueberries any more because of tapeworms. The raccoon dog is Asian and belongs there, not here. Similarly with the mink; it was introduced in 1930, but now we shall kill it. The black headed gull, on the other hand, has flown here on its own wings, therefore it may stay.

Q: How long must you have been here, before you are accepted? A: The mink is alien, and will never be accepted.

Q: But the raccoon dog has walked on its own legs from Russia? A: Yes, but it was transported from Korea to Russia. Had it walked all the way by itself, it would have been different.

Q: Why are we spending large amounts of money on reintroducing the wolf – which eats us – while we shall exterminate the raccoon dog which just plays dead when threatened? A: This is a political decision.

Q: Do you hate raccoon dogs? A: This is not about emotions, but about scientific judgment.

Q: But still, why will you reintroduce wolves but exterminate the raccoon dog? A: It is a question of value based choice..

Q: What kind of values? A: What we want with Norwegian nature...

This case was chosen because it reveals typical human concerns as they shift between rights and consequences. We can be emotionally swayed by questions of aliens' rights, as well as by a sense of duty to preserve or restore the environment to some pristine state. That such attitudes are common can be documented by the reactions to the article, "Don't judge species on their origins" (Davis et al., 2011). The authors clearly hit a nerve with an amazing 10 400 hits on a Google search (24.08.2011) with the full title. Most comments seem to be supportive of the article, but there are also a number that are negative, such as Hough Snee of *Perceptible Changes* (2011). Davis et al. observe that "'non-native' species have been vilified for driving beloved 'native' species to extinction and generally polluting 'natural' environments. Intentionally or not, such characterizations have helped to create a pervasive bias against alien species that has been embraced by the public, conservationists, land managers and policy-makers, as well as by many scientists, throughout the world." Their main point is that management of introduced species should be based on rational, not emotive reasons. Vince (2011) has a similar opinion based on experience from the Galápagos, where eradication programs of invasive plants like blackberries have proven futile. But on the other hand, the resulting hybrid ecosystem turned actually out to be acceptable and could even be "worthy of conservation". That a species is alien is actually a poor predictor of its environmental impacts, which can be detrimental as well as beneficial.

The issue of being native or alien was first introduced by the English botanist John Henslow in 1835. However, it was not until the 1990's that it became a global public policy to try and preserve pristine environments by eradicating aliens (Wittenberg and Cock, 2001) since they were considered to be a leading threat to biodiversity and a cost to human enterprises, as well as a threat to health.

6. Ethical theories

The three classical ethical mindsets of consequentialism, duty ethics, and virtue ethics give different reasons for choice (Blackburn, 1998). According to consequentialism, an action is morally good if the intended consequences are good. Consequentialism thus makes the good prior to the right, and it defines the right operation in terms of promoting the good. Thus, a consequentialist looks neither at the nature of the action itself, nor at the character or attitude of the decision-maker: only consequences count. This contrasts with Kantian duty ethics that defines the right prior to the good. The principle of morality according to Immanuel Kant is to act only on that maxim through which you at the same time will that should become a universal law. It considers whether the decision-maker has obeyed the right principles, and thereby fulfilled her duty or obligations, no matter what the consequences are. Finally, virtue ethics is only concerned with the character and attitude of the decision-maker; an action is morally right if the relevant virtues have been displayed, such as courage, loyalty etc. To be principled is a virtue as well, and this provides a link between duty and virtue ethics: to fail at duty ethics is to fail at virtue ethics (Wenstøp and Koppang, 2009). The reason I call the three ethical theories 'mindsets' is that people are often unaware of their own reasons for choice, even though an ethical mindset pervades their emotions and thinking. Making people aware of this would provide for better mutual understanding in value conflicts (Wenstøp, 2005).

7. Classification of arguments

The dialogue in the case has the appearance of a bewildering – sometimes contradictory – array of arguments and attitudes, which may become clearer if one could classify them according to reasons for choice. Thus, the three classical ethical theories of virtues, duties and consequences are natural candidates.

One class of arguments in the case describes the character of the alien; the raccoon dog is portrayed as evil. While this is sometimes said to vilify human immigrants, it is questionable when used against animals. Is a cat playing with a mouse evil, or just inquisitive? Still, it is natural to classify such arguments under the label "virtue ethics" since virtue has to do with the character of the agent, in this case the raccoon dog. And accusations of want of virtue are bound to elicit strong negative emotions such as disgust, anger or xenophobia.

The next class of arguments is about rights. Who has the right to live in Norway? The argument goes that you are okay if you have arrived by walking, but you must walk (or fly) the whole way; half is not enough. As rights have to do with laws and rules, I propose to classify these arguments under duty ethics. The question of right to land is of course problematic, be it human immigrants, plants or animals. But regardless of who the transgressor is, violation of perceived rights generally elicits very strong negative emotions. Baron and Spranca (1997) have introduced the term 'protected values' for values that are protected by rules or rights. The dialogue also has an undertone of a sense of duty to preserve or restore a pristine environment; an attitude which is prevalent among lay people and conversationalists alike.

The third class of arguments has to do with what we want with Norwegian nature. In other words: what would be the consequences if we welcomed the raccoon dog, and do we like those consequences? Such issues belong to consequentialism; the value of an action depends solely on its consequences. This is the third ethical category and it is interesting to note that the answer requires for and against judgments, where the raccoon dog's character weighs in on the scales.

Consequentialism appears to be the most common attitude among environmental scientists, while duty ethics often underpin the attitudes of managers and policy-makers. But sometimes there is a hierarchy of ethical platforms: According to the authoritative "Toolkit of Best Prevention and Management Practices" for control of "Invasive Alien Species" (Wittenberg and Cock, 2001), "the ultimate goal of the strategy should be preservation or restoration of healthy ecosystems". Thus there is a duty to preserve or restore, but then one has to define what healthy means. This brings us into the realm of consequentialism where one needs to identify criteria for healthiness: "Thus, the initial step in a national programme must be to distinguish the harmful from the harmless alien species and identify the impacts of the former on native biodiversity." (Ibid.)

8. Affect and deliberation

We have already identified emotions as a factor at work in environmental discussions. To understand the deeper connection between ethics and emotions, it is useful to consider the emerging field of neuroeconomics, which studies neural correlates of economic decision-making (Camerer et al., 2005). Neuroscientists use several techniques, such as positron emission topography (PET) scanning and functional magnetic resonance imaging (fMRI), to monitor the location and pattern of neural activity in the brain when decisions are made. They have established that there are two kinds of neural processes involved in decision-making: cognitive and affective. This comes as no surprise; Plato characterized human behavior as riding in a chariot drawn by two horses, reason and passion. What is new is that we can now observe that humans actually are hard-wired in that way, and that the processes often compete for dominance. Passion is quick in the onset, sometimes evoking immediate action before the slower deliberate processes can become engaged and hinder an unreasonable response.

It is interesting to note the fields of law and economics have different traditions in this respect. The concept of *homo economicus* or 'economic man' is strong in economics. Here, passions are kept at arm's length, and it is assumed that self-interested actors have the ability to make deliberate judgments toward their subjectively defined ends. Law, however, incorporates a notion of passion exemplified by the legal maxim *Ira furor brevis est* (anger is short insanity) and this is occasionally used to excuse an offender.

8.1 Emotions

Neuroscientists have demonstrated that affective states have somatic correlates, i.e. emotions that work together or in competition with reasoning processes to shape decision making. Thus, neuroeconomics seeks to bring passion back into economic models to build more complete models of human decision processes. Emotions need not be consciously felt, but almost all actions seem to be prompted by emotions. They work to improve our affective state by giving the body appropriate response signals. The interplay between affective processes in our brain and emotions in our body is massively parallel, with many pathways working simultaneously and rapidly, supporting the observation that we sometimes act before we have time to think.

8.2 Conation

According to ancient wisdom, *Conatus* is one of three parts of the mind, along with the affective and cognitive. While feelings come from the affective system and thoughts from the cognitive system, the conative system drives how one acts on those thoughts and feelings. These classical concepts where known to Aristotle and are congruent with another important observation in neuroscience: that cognitive processes alone cannot produce action. Conation – the desire to act – requires that the cognitive system works through the affective system. Any action is preceded by an emotion. The picture is therefore that although the cognitive system is used for searching for options and predicting consequences of actions, it cannot evaluate those consequences. That must be done by the affective system. Damasio (1994) made notable empirical observations in neuroscience when he discovered that people with damage to the prefrontal lobes were emotionally flat when they contemplated future consequences of decisions. As a consequence, they were very poor decision-makers, sometimes being completely unable to make decision, sometimes making decisions that were obviously detrimental to their own well-being. These people had severed the connection between the cognitive system and amygdala, which is responsible for eliciting somatic responses. According to Damasio, it is the feeling of these emotions that prompt action. In the words of Camerer et al. (2005) "It is not enough to "know" what should be done; it is also necessary to "feel" it". Figure 1 shows a model of the two processes of affect and deliberation involved in conation.

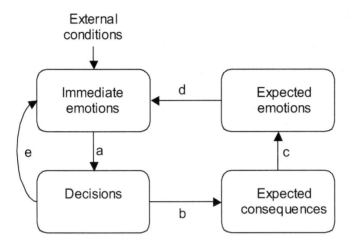

Fig. 1. A model of decision-making between affect and deliberation, adapted from Loewenstein and Lerner (2002). Path a: all decisions are prompted by immediate emotions. The cognitive pathway a-b-c-d is employed in deliberation, which involves prediction of consequences of alternative decisions and how good they will feel. The expected emotions influence immediate emotions, which in turn may prompt action. Pathway e is the affective pathway caused, for example, by fear or disgust that may affect immediate emotions more strongly than expected emotions.

Figure 1 illustrates the two – often competing – processes involved in conation: The affective pathway a-e-a, and the deliberate pathway a-b-c-d-a. Both pathways involve emotions, the difference being that they are fast and strong in the affective pathway and slow and temperate in the deliberate one. The affective pathway, however, does not involve cognitive processes; here one acts without thinking.

With this background, it is easier to understand the mechanisms behind conflicts in environmental management. Conflicts are created by the opposing forces of strong affect in some stakeholders and tempered emotions in another stakeholder. For example, pathway e represents one stakeholder's affect resulting from the perceived virtues of animals – be they bad (raccoon dogs) or good (whales). Another stakeholder may have a more balanced emotional response elicited through rational deliberation about consequences of actions, using the pathway a-b-c-d. Such conflicts can even be intra-personal – as when a person is of different minds – not only interpersonal, as when stakeholders argue from the vantage points of divergent mindsets.

8.3 Rationality

If we want to define rationality in a way that makes rationality bode well for decision-making, it is important to note that the cognitive pathway in Figure 1 involves emotion, suggesting that a good definition of rationality should incorporate it. Since pure thinking is not sufficient to prompt action, any concept of rationality that does not incorporate emotion would be insufficient. Rationality requires that deliberation about consequences be infused with emotion (pathway b-c-d), while one avoids strong affect (pathway e) that precludes thinking. Therefore, a rational decision-maker should be conscious that good decision-making requires temperate emotions that stir emotions which will then enact decisions according to the results of deliberation. Incidentally, the word 'deliberation' means literally to balance or weigh, and is derived from Latin *libra*, meaning 'scale'.

There are many definitions of rationality available, most of them connecting rationality with reason, but we need a definition that incorporates values and lends itself to inclusion of emotion. Føllesdal's definition of (1982) provides a useful starting point. He defines rationality in four dimensions:

1. *Rationality as logical consistency.* This pertains to values as well as beliefs, and is the central pieces of classical notions of rationality
2. *Rationality as well-foundedness of beliefs.* This means that beliefs about facts are well supported by available evidence, and that one has made a decent effort at securing relevant information.
3. *Rationality as well-foundedness of values.* One should have obtained reflective equilibrium that gives a stable set of convictions that are relevant for the decision situation.
4. *Rationality of action.* In practice, this means application of decision theory including maximization of expected utility.

While Føllesdal and others believe that, based on the principles above, one can use reason to determine which decision is rational, the findings in neuroscience suggest otherwise: conation — the desire to act – also requires emotion. One has to feel in order to act. We thus need to revise the fourth dimension:

4. *Rationality of conation*: Elicitation of tempered emotions that enact the beliefs and values.

The point here is that beliefs about consequences are not enough; one must also have a feeling for them. This is the only way we can ensure an ethical consequential decision-making.

9. Ethics, values and emotion

This paper has sketched a picture of a conspicuous consilience among the theories of value, ethics and neuroscience, which was first noted by Wenstøp and Myrmel (2006). We used a dialogue about control of invasive species as a background, but let us now complete the picture, first by using a general organization as an example in this chapter, and then by applying it to environmental management in chapter 10.

First, the character of people can be described in terms of virtues. There are several virtue systems such as the four cardinal virtues (prudence, justice, restraint, and courage) and the seven heavenly virtues (chastity, temperance, charity, diligence, patience, kindness, and humility). Table 1 shows the most popular corporate core values, which are models for peoples' behavior in companies. Virtuous people are met with positive emotions, but if, on the other hand, they display a lack of virtue, negative responses are usually swift and strong. Thus, there is a correspondence between virtue, and thereby virtue ethics, and affective responses through pathway e in Figure 1, which bypasses the cognitive system in people and elicits strong emotions.

Value category	Ethical theory	Emotions	Value examples
Virtues	Virtue ethics	Strong	integrity, honesty, respect, openness, fairness, innovativeness, trustworthiness, creativeness, reliability, dignity
Protected values	Duty ethics	None	Values protected by voluntary standards and certificates
Created values	Consequentialism	Tempered	Stakeholder values such as return on equity, work places, products and services

Table 1. Correspondence of value category, ethical theory and emotion elicited in decision-making. Typical organizational values are shown as examples.

Second, important values are often protected by law, rights or custom; the right to land is a notable example with environmental implications. Corporations, for example, sometimes subscribe voluntarily to standards or certificates such as the UN's human rights charter against the use of child labor, and international industry standards against pollution, etc. Following such guidelines requires no emotion, but transgression is bound to elicit strong emotions.

Third, the intended consequence of organizational actions is creation of value, and in general we talk about creating value for the stakeholders. However, stakeholder values are often in conflict and decision-making requires making trade-offs among them. This again requires temperate emotions through the cognitive pathway a-b-c-d in Figure 1. See the third line in Table 1.

10. Application to environmental management

In their paper on value conflicts in OR, Wenstøp and Koppang (2009) propose to benchmark conflict potential according to two dimensions: The degree to which the decision criteria represent intrinsic rather than instrumental values, and the extent of stakeholder involvement. They assume that decision criteria which are only technical means to further ends are less likely to create conflicts than if they represent ends that people easily attach value to. Further, it makes a difference whether the decision is made on behalf of people – such as in a board room, or whether stakeholders participate in the process, which makes the conflict potential higher.

Conflicts in environmental management notoriously engage many stakeholders who will differ over intrinsic values, and this therefore makes the conflict potential high. The raccoon dog is perhaps a minor threat, but it managed to call the attention of the Directorate for Nature Management, ornithologists, hunters, conversationalists, journalists, cabin owners, animal protectionists etc. Such conflicts call for a conceptual basis that makes it possible to understand peoples' reasons and the sources and nature of their emotions. Such a conceptual basis was outlined in the previous chapter, and when it is applied to environmental management, it takes the form shown in Table 2.

Value category	Ethical theory	Emotions	Examples
Virtues	Virtue ethics	Strong	Character and intelligence of animals, character of human agents
Protected values	Duty ethics	None	Endangered species, special biotopes, animal rights
Created values	Consequentialism	Tempered	Biodiversity, recreation, beauty, economic resources, food

Table 2. Value categories with examples of values in environmental management, the underlying ethical theory, and the level of emotions elicited in decision-making contexts.

10.1 Virtues

At the end of the sixteenth century, Michel de Montaigne claimed that animals are both moral and rational, but it was not until the seventeenth century that the debate gained widespread attention (Harrison, 1998). Such beliefs are still with us today as is evident when protectionists claim that animals are virtuous, or the opposite. We have seen how raccoon dogs have been vilified. Whales represent an opposite example; they have been sanctified and any discussion about whaling notoriously stir very strong emotions (Reiss, 2008). Such claims defy rationality and as such are not within the domain of OR techniques.

Thus, virtue values are outside the scope of rational approaches. That does not mean that they are irrelevant, however. All reasons are relevant but it should be recognized that the virtues of a species as a reason for decision must be treated differently from consequential reasons since they cannot be traded off on the same scale.

10.2 Protected values

Some values are regarded as being sufficiently important to be protected through laws, rules or regulations; and obeying them need not involve emotions – only the coldness of a bureaucratic heart, or as Weber (1947) put it: "The dominance of a spirit of formalistic impersonality, sine ira et studio, without hatred or passion, and hence without affection or enthusiasm". OR needs only to take such rules as frames or restrictions.

The problem is, however, that some values reach a protected status through emotional processes that are not necessarily rational. In the case of invasive species, xenophobia is rampant. The question of the rights of aliens, for instance, is provocative, since it is impossible to make consistent rules. The simple question, "How long must you be here to become native?" defies any logical answer. And since the real reason for pointing the finger at alien invasives is that they pose a major threat to biodiversity, the issue should become consequentialistic and not a question of rights; emotions should thus be tempered accordingly.

The same can be said about the xenophobia that is created when alien plants or trees are considered a threat to the native landscape. According to Olwig (2003), this notion stems from a particular post-Renaissance concept of landscape, space and nature that ultimately derives from what he calls a 'cartographic–pictographic episteme'. Instead of trying to protect the existing landscape, we should acknowledge that landscapes do change and rather ask what kind of landscape do we want.

Another dubious protected value is 'genetic integrity' (Smout, 2003), which has led to campaigns against species introductions that might interbreed with natives. A more defensible approach, according to Smout, might be to revive the notion of some species as pests, but to hesitate before involving conservation in anything analogous to ethnic cleansing for other species.

10.3 Created values

The consequences of environmental management are called created values in Figure 1. These are the end impacts of actions, and OR embraces a set of tools that is well suited to develop consequence models to predict end impacts. System Dynamics is one example from this toolbox. But since any action is bound to have several impacts, one also needs to weigh them according to importance, for example, how much do we prefer ducks over raccoon dogs (or is it the other way around)? For that purpose we need to elicit temperate emotions among stakeholder and multi criteria methods may be useful here. See Seip and Wenstøp (2006) for an overview.

Thus, the basic recommendation of this paper is to look at the values and the reasons for them first. Put virtues aside as they must be addressed through processes outside OR. Question the protected values because they may be consequence values in disguise, and address the created values by conventional OR methods.

10.4 Illustration: Peter Singer on whaling

The question of whether or not to allow whaling has long been on the international forum for environmental management controversies, and few debates are more heated, with traditional whaling countries like Japan, Iceland and Norway stand on one side of the issue and environmentalists and ethicists on the other. The International Whaling Commission (IWC), with 89 member countries, is a central actor with a main duty to keep under review and revise the measures that govern the conduct of whaling throughout the world. These activities include protecting certain species, designating whale sanctuaries and setting limits on the size of catches. While the IWC's agenda is primarily scientific, based on a consequentialistic approach, any member country can reserve itself from decisions that IWC makes, and such reservations are usually made on emotional grounds.

Peter Singer is a well known ethicist and spokesman for animal rights, including whales. Let us see what he has to say in this connection (Singer, 2008): "I did not argue that whaling should stop because whales are endangered". But "whales are social mammals with big brains, capable of enjoying life and of feeling pain – and not only physical pain, but very likely also distress at the loss of one of their group." He further argues that whales cannot be humanely killed, they are too big, and using explosives would mean loss of flesh and oil, which is the very reason for hunting whales. "So harpooned whales typically die slowly and painfully." He concludes that "Causing suffering to innocent beings without an extremely weighty reason for doing so is wrong. If there were some life-or-death need that humans could meet only by killing whales, perhaps the ethical case against it could be countered. But there is no essential human need that requires us to kill whales. Everything we get from whales can be obtained without cruelty elsewhere. Thus, whaling is unethical."

We see that Singer starts by laying aside the consequential issue of whether whaling is sustainable. This is a scientific issue and on the agenda of the International Whaling Commission. Instead, he turns to more emotional issues, first by attributing virtue to whales, and then by arguing that whales should be protected because of their size – they cannot be killed without suffering. He does concede, however, the possibility of a consequential trade-off here, but after inspecting the ethical scales, he concludes that whales should be protected.

Thus Singer visits all three ethical categories in his chain of arguments, stirring emotions by attributing virtue to whales, as well as to convince the reader that whales should be protected to avoid suffering. Regardless of the outcome of a scientific consequential analysis of the pros and cons of whaling – such as sustainability against flesh and oil, he concludes that whaling is unethical.

Singer goes on to argue against Japan's attitude. They say "that it [Japan] wants the discussion of whaling to be carried out calmly, on the basis of scientific evidence, without "emotion." The Japanese think that humpback whale numbers have increased sufficiently for the killing of 50 to pose no danger to the species. On this narrow point, they might be right. But no amount of science can tell us whether or not to kill whales." He then dismisses Japan's call for an emotionless, scientific evaluation, seeing little added value for the Japanese regarding nutrition and health, and then accuses the Japanese of being emotional themselves, since the real reason for whaling seems to be to protect the whaling tradition.

Singer's arguments are well structured, and it is easy to identify his ethical platforms as described in this paper. They are overwhelmingly emotional, however, and therefore not susceptible to rational arguments that might be raised from a scientific OR point of view. Any decision-maker is therefore left to consider all reasons for and against whaling and use his own judgment in the matter.

10.5 Rational consequentialistic analysis

In management of operations, ethical decision making should start by separating created values from protected values and then proceed to work with the created ones. Protected values are highly emotional and not amenable to rational trade-off analysis, while created values can be handled through emotionally tempered processes such as multi criteria decision analysis (MCDA). See Belton and Stewart (2002) for a thorough presentation of MCDA methods. In general, the process runs like this: First one needs to represent the created values with quantitative measures, which are called decision criteria (x_1, x_2,..). Then one estimates the consequences of the decision alternatives (A_1, A_2, ...) in terms of decision criteria scores. Uncertainty can be represented by probabilities or handled through scenario analysis. Since the consequences generally are measured on different scales, it is necessary to bring them onto the same scale, which can be done with value- or utility functions. The advantage of utility functions is that they represent attitudes towards risk; the disadvantage is that they are more demanding to assess. A decision problem with three alternatives and three decision criteria would be described by a table like in Table 3.

	Option 1: A_1	Option 2: A_2	Option 3: A_3	Weight
Criterion 1	$x_1(A_1) \rightarrow u_{11}$	$x_1(A_2) \rightarrow u_{12}$	$x_1(A_3) \rightarrow u_{13}$	w_1
Criterion 2	$x_2(A_1) \rightarrow u_{21}$	$x_2(A_2) \rightarrow u_{22}$	$x_2(A_3) \rightarrow u_{23}$	w_2
Criterion 3	$x_3(A_1) \rightarrow u_{31}$	$x_3(A_2) \rightarrow u_{32}$	$x_3(A_3) \rightarrow u_{33}$	w_3
Utility	u_1	u_2	u_3	

Table 3. An MCDA decision table with three alternatives and three decision criteria.

The overall utility of an alternative is usually calculated as the weighted sum of utilities: $u_1 = w_1u_{11} + w_2u_{21} + w_3u_{31}$, etc. although more complicated functions that include synergy effects among the variables are available (Keeney and Raiffa, 1976) .

In cases with real value conflicts, no decision alternative will dominate the others in the sense that it scores better on all criteria, and then the optimal decision will necessarily depend on the importance of the criteria, which are represented by weights in Table 3. While the scores are beliefs about real consequences, the weights are intrinsically subjective and will therefore depend on the values of the decision-maker. This creates two challenges. First, for a given decision-maker, one need to obtain the weights with methods that elicits temperate emotions through vivid rendering of future scenarios (Wenstøp, 2005). The field of MCDA offer several methods for achieving this (Belton and Stewart, 2002). Second, in environmental management there will usually be many stakeholders with different values, and one way to take these into account, is to try to identify viable compromises through suitable processes (Wenstøp and Koppang, 2009).

Let us now return to the issue of whaling. The two main arguments against whaling in the public debate are: (1) it is cruel (Singer, 2008), and (2) whales have rights (Johansen, 2005). Singer, as we have seen, stirs emotions in the way he argues that whales cannot possibly be killed in a humane way. Against this, the Japanese argue that, yes, by using the electric lance, whales can be killed in a humane way, at least if one compares the time it takes before the whale dies to what happens in big game hunting (Hayashi, 1996). Thus Hayashi argues that whales should not be protected by the humane killing argument, but that one should rather allow for trade-offs and treat humane killing as a created value, which then could be measured in terms of survival time in the killing process, which should be as short as possible.

The debate concerning man's rights versus animal's rights is less amenable to rational analysis. From one side the whale is portrayed as a "symbol of the mighty, uncorrupted and innocent nature as compared to the greedy, revengeful and morally depraved man". From the other side, man is portrayed as a steward on earth: "When man ate of the tree of knowledge, lost his innocence and left paradise there was no way back. Everywhere where man went to live he formed the vegetation and the landscape as a consequence of his use of nature. Man found his place in competition with and at the sacrifice of other species. He crowded out wild animals when they were competitors for food and tamed others as working force or used them as producers of food. This was, and still is, a prerequisite for population growth, increased productivity and cultural development. Taming of animals, use of animals, and killing of animals for food are indispensable and necessary prerequisites for man to be man; that is to build civilizations."(Johansen, 2005). – It is fair to say that this emotional debate defies consequentialistic rationality, and must be fought in a different arena.

Having thus separated created from protected values, one can proceed with those that are perceived as created (or destroyed) by whaling and which are amenable to rational analysis and trade-offs. Among these are:

-    Sustenance of aborigine populations, measured as the size of populations sustained by whaling.
-    Sustenance of costal populations, measured in terms of annual income from whaling.
-    Health improvement from diet based on marine fatty acids, measured in terms of life years.
-    Scientific information about ecosystems, especially fish/whale interactions, using the number of whales killed as an indicator.
-    Commercial hunting, measured in terms of profit
-    Suffering of whales killed, measured in terms of time spent in agony.
-    Sustainability of whale stock, the size of the stock used as an indicator.

These created values and others can form the basis of a rational, emotionally tempered, analysis of decisions, such as setting quotas for particular whale species. This would involve estimation of consequences and subjective weighting, thus producing data that would enter a table like Table 3.

11. Conclusion

This paper has a modest aim: to argue that conflicts in environmental management can be better understood by sorting out the arguments according to the underling ethical platform. This provides an understanding of the degree of emotions involved and this platform serves as a tool for identifying those values that are consequential and therefore amenable to rational trade-off analysis. OR's proper arena is to provide facts regarding the consequential values and to assist in making balanced trade-offs among them. Within its proper domain, OR cannot deal with emotion-laden values such as virtues, although processes outside OR may be useful.

12. Acknowledgment

I want to thank Carl Brønn and Søren Wenstøp for valuable discussions and help with the manuscript.

13. References

2008. Handlingsplan mot mårhund, Nyctereutes procyonoides. Norway: Directorate for Nature Management.

Baron, J. & Spranca, M. 1997. Protected values. *Organizational Behavior & Human Decision Processes*, 70, 1-16.

Belton, V. & Stewart, T. J. 2002. *Multiple Criteria Decision Analysis,* Dordrecht.

Blackburn, S. 1998. *Ruling Passions, A Theory of Practical Reasoning,* Oxford, Clarendon.

Boulding, K. E. 1966. The ethics of rational decision. *Management Science,* 12, B 161-169.

Brans, J.-P. 2004. The management of the future: Ethics in OR: Respect, multicriteria management, happiness. *European Journal of Operational Research,* 153, 466-467.

Brans, J.-P. & Gallo, G. 2007. Ethics in OR/MS: past, present and future. *Annals of Operations Reseach,* 153, 165-178.

Brans, J. P. & Kunsch, P. L. 2010. Ethics in OR and sustainable development. *International Transactions in Operational Research.*

Camerer, C., Loewenstein, G. & Prelec, D. 2005. Neuroeconomics: How neuroscience can inform economics. *Journal of Economic Literature,* XLIII, 9-64.

Damasio, A. R. 1994. *Descartes' error. Emotion, Reason and the Human Brain,* New York, G P Putnam's sons.

Davis, M. A., Chew, M. K., Hobbs, R. J., Lugo, A. E., Ewel, J. J., Vermeij, G. J., James H. Brown, Rosenzweig, M. L., Mark R. Gardener, Carroll, S. P., Ken Thompson, Pickett, S. T. A., Stromberg, J. C., Tredici, P. D., Suding, K. N., Ehrenfeld, J. G., J. Philip Grime, Mascaro, J. & Briggs., J. C. 2011. Don't judge species on their origins. *Nature,* 474, 153-154.

Føllesdal, D. 1982. The Status of Rationality Assumptions in Interpretation and in the Explanation of Action. *Dialectica,* 36, 301-316.

Harrison, P. 1998. The Virtues of Animals in Seventeenth-Century Thought. *Journal of the History of Ideas,* 59, 463-484.

Hayashi, Y. 1996. *Humane Killing of Whales and the Sustainable Wildlife Utilisation* [Online]. Available: http://luna.pos.to/whale/icr_21_haya.html [Accessed 4. Nov. 2011].

Hough-Snee, N. 2011. *Article alert: Don't judge species on their origins* [Online]. Perceptible changes. Available: http://perceptiblechanges.blogspot.com/2011/06/article-alert-dont-judge-species-on.html [Accessed August 25 2011].

Johansen, H. P. 2005. Opposition to Whaling - Arguments and Ethics. *In:* ENVIRONMENT, M. R. A. (ed.). Oslo: Norwegian Ministry of Fisheries and Coastal Affairs.

Keeney, R. & Raiffa, H. 1976. *Decision with Multiple Objectives,* New York, John Wiley & Sons.

Kunsch, P. L., Kavathatzopoulos, I. & Rauschmayer, F. 2009. Modelling complex ethical decision problems with operations research. *Omega,* 37, 1100-1108.

Le Menestrel, M. & Van Wassenhove, L. N. 2004. Ethics outside, within, or beyond OR models? *European Journal of Operational Research,* 153, 477-484.

Le Menestrel, M. & Van Wassenhove, L. N. 2009. Ethics in Operations Research and Management Sciences: A never-ending effort to combine rigor and passion. *Omega,* 37, 1030-1043.

Loewenstein, G. & Lerner, J. S. 2002. The role of affect in decision making. *In:* DAVIDSON, R. J. (ed.) *Handbook of Affective Sciences.* Cary, NC, USA: Oxford Universiry Press Inc.

Meadows, D. H., Meadows, D. & Randers, J. 1972. *The Limits of Growth. A Report for The Club of Rome's Project on the Predicament of Mankind,* New York, Universe Books.

Olwig, K. R. 2003. Natives and Aliens in the National Landscape. *Landscape research,* 28, 61-74.

Rauschmayer, F. 2001. Reflections on ethics and MCA in environmental decisions. *Journal of Multi-Criteria Decision Analysis,* 10, 65-74.

Reiss, M. 2008. *Ethics of whaling* [Online]. Waikato: Biotechnology Learning Hub, The University of Waikato. Available: http://www.biotechlearn.org.nz/themes/bioethics/ethics_of_whaling [Accessed September 4 2011].

Rosenhead, J. 2005. Problem structuring methods as an aid to multiple-stakeholder evaluation. *In:* MILLER, D. & POTASSINI, D. (eds.) *Beyond benefit cost analysis: accounting for non-market values in planning evaluation.* Aldershot, UK: Ashgate Publishing Ltd.

Seip, K. L. & Wenstøp, F. 2006. *A Primer on Environmental Decision-Making: An Integrative Quantitative Approach*, Springer Verlag, Dordrecht.

Singer, P. 2008. *Hypocrisy on the High Seas?* [Online]. Project Syndicate. Available: http://www.project-syndicate.org/commentary/singer32/English [Accessed September 5th 2011].

Smout, T. C. 2003. The Alien Species in 20th-century Britain: constructing a new vermin. *Landscape research,* 28, 11-20.

Sætre, S. 2010. Den fremmede (The alien). *Morgenbladet,* 15-20 October, p.8.

Vince, G. 2011. Embracing invasives. *Science,* 331 March 18, 1383-1384.

Weber, M. 1947. *TheTheory of Social and Economic Organisation,* New York, Free Press.

Wenstøp, F. 2005. Mindsets, rationality and emotion in Multi-criteria Decision Analysis. *Journal of Multi-Criteria Decision Analysis,* 13, 161-172.

Wenstøp, F. 2010. Operations research and ethics: development trends 1966–2009. *International Transactions in Operational Research,* 17, 413-426.

Wenstøp, F. & Koppang, H. 2009. On operations research and value conflicts. *Omega,* 37, 1109 - 1120.

Wenstøp, F. & Myrmel, A. 2006. Structuring organizational value statements *Management Research News,* 29, 673 - 683.

Wilson, E. O. 1998. *Consilience: The Unity of Knowledge,* New York, Alfred A. Knopf, Inc.

Wittenberg, R. & Cock, M. J. W. (eds.) 2001. *Invasive Alien Species: A Toolkit of Best Prevention and Management Practices,* Wallingford, Oxon, UK: CAB International.

6

Harnessing Efficiency and Building Effectiveness in the Tax Department

Yair Holtzman and Laura Wells
Director WTP Advisors, Business Advisory Services Practice Leader
USA

1. Introduction

Organizations employ a number of formulas to improve their business processes. These actions typically involve searching for internal cost-savings opportunities, developing departmental strategic relevance and efficiencies, and demonstrating an enhanced focus on increasing profitability. The financial crisis that began in 2007 continues to put stress on companies in the global marketplace. As a result, these companies face extreme pressure to tighten budgets and improve their practices. Nevertheless, the tax function is rarely taken into consideration as a viable arena through which to operate more profitably when organizations look to improve the efficiency and effectiveness of their internal operations. This omission can largely be attributed to the perception of the tax department held by the majority of business executives. That is, the tax department is typically viewed strictly as a cost center, a necessary component for the continued functioning of the organization but unable in itself to generate any revenue. This cost center mentality does not need to be the case. The tax department has the potential to produce data analysis regarding future periods rather than simply reporting on the historical data of the company. This perspective enables the tax department to take on an active role in contributing to the organization's strategic, forward-looking directives. The selective implementation and integration of operations strategy tools and methods, with a focus on maximizing process improvement and efficiency enhancement, can lead to increased value within the tax department. Further, these management and strategy practices will expand the role the tax function plays in facilitating business process improvements, thus leading to increased value within the organization as a whole.

2. The current state of the tax department

A tax department holds four primary responsibilities: compliance activity, tax accounting, risk mitigation, and strategic tax planning. The tax department's domain historically has been a highly specialized and discrete world in which professionals focus primarily on fulfilling a complex web of tax requirements from federal, state, and foreign tax authorities. As a result, complying with legislation has dominated the tax department's agenda in recent years. Surveys indicate that tax professionals spend approximately 84 percent of their time on compliance activity. This "low value added" data management and compliance work inundates many tax personnel preventing them from contributing to strategic organizational goals. Contributing to strategic organizational objectives requires that the appropriate tax employees have the capacity to perform forward looking and "high value added" analysis including strategic planning activity as part of the leadership team. Assume this figure of 84 percent is correct and that these individuals work a 40 hour work week. Cutting the time spent on compliance activity by a conservative 25 percent would provide each individual with the opportunity to expend between 400 and 500 additional hours per year on higher value added activity.

The increased examination of the tax function by shareholders, the audit committee, and the executive suite, in conjunction with increased regulatory requirements, is trapping the tax department in a low value loop, thus preventing the department from achieving its potential to contribute strategic value to the organization. This low value loop results as companies attempt to collect and report information in accordance with tightened governmental control.

The prevalence of increased governmental involvement is demonstrated through regulations such as the Sarbanes-Oxley Act of 2002. The Sarbanes-Oxley Act (HR 4173) set new standards for the corporate management and accounting processes of publicly traded companies, and it requires strict documentation of business process controls affecting financial reporting. Section 404(b) is considered one of the most pertinent to compliance because it requires companies to report on the managerial assessment of their internal controls. Sarbanes-Oxley-related compliance initiatives play a major role in driving change within documentation, compliance, and reporting measures of the tax function in recent years.

Additionally, the Federal Accounting Standards Board, through measures such as FAS 109 and FIN 48, has involved itself in working to create more transparency. The Board determined that "financial statements should reflect the current and deferred tax consequences of all events that have been recognized in the financial statements or tax returns" (63). FAS 109 requires companies to establish deferred taxes for temporary, or timing, differences in their assets and liabilities. That is, as defined in the FAS 109 objectives, companies must account for differences "between the tax basis of an asset or a liability and its reported amount in the statement of financial position [that] will result in taxable or deductible amounts in some future year(s) when the reported amounts of assets are recovered and the reported amounts of liabilities are settled." This measure affects the amount of disclosure required of companies.

Legislative changes and updated regulations affecting tax department responsibilities continue to transpire. More recently, in 2010, the IRS released Schedule UTP (Uncertain Tax Positions), requiring designated businesses to report uncertain tax positions on their tax returns. Schedule UTP is being phased-in over a five year period. This has allowed many tax departments to delay addressing the issue.

Amidst their efforts to comply with these legislative measures, many companies have found their tax processes to be poorly controlled. Several factors play into these poor controls in addition to legislative measures. The tax function adheres to calendars and timetables unique from other departments, and it uses different processes and systems, accounting methods, and reporting standards than the rest of the finance function or the business at large. Studies have identified that one third of material weaknesses in controls were tied to tax accounting. These weak organizational practices play into the inefficiencies of the tax department and highlight an area of immense opportunity for improvement through internal transparency and integration of higher quality control. The poor controls currently in place and the tightened regulations being enacted demonstrate a significant need within the tax department for the production of consistent information and data that is reliable, organized, and coherent.

Growth and restructuring strategies often result in disorganized and inconsistent production of information, particularly in the area of mergers and acquisitions. Various tax personnel, software systems, and compliance processes are also merged and result in this disorganization and inconsistency. Mergers and acquisitions can lead to each location or business entity reporting using different accounting systems, formatting, or even interpretation of accounting rules. Newly acquired information management systems are often incongruous with a company's established systems. The use of varied systems results in inconsistent data, and often considerable manual work is required to standardize, consolidate, and integrate the data. This incongruity typically results in a conglomeration of manual methods and complex spreadsheets. As a result, accounting and tax personnel are spending an inordinate amount of time and effort collecting, validating, and manipulating data, and they are unable to contribute to the profitability of the company through higher value added activities.

Furthermore, the prior system often continues to be run concurrently with the new system when new software is implemented. This practice is beneficial in the short term, but the duration of this concurrent operation takes place much longer than is necessary for implementation purposes. Delaying the completion of the transition adds additional steps rather than simplifying the process. Compiling data for reporting cycles in multiple systems becomes an extremely complex, laborious process.

Many companies report for management and tax purposes on different bases, whether that is by business unit or legal entity. This split in reporting further complicates the procurement of quality tax information. The two sets of information do not always reconcile easily. Handling data procurement issues and irreconcilable reports leaves little opportunity for the tax department to become deeply involved in higher level analysis and high value added activities such as tax planning and strategizing. Accordingly, tax is disconnected from contributing to fundamental corporate strategies or forward-looking directives. For obvious reasons, the absence of tax professionals in the strategic decision making process is detrimental not only to the department but also to the company as a whole.

The integration of tax accounting systems and financial reporting systems enables tax to take on a more active role in organizational planning and strategy. An enterprise needs systems in place that produce the data and information required for tax planning, tax compliance, and developing and evaluating tax strategy. In most companies, numerous systems feed into a core general ledger, and no single integrated system can easily produce data for all required finance, accounting, and tax purposes. The resulting process inefficiency leads to data complexity, manual procedures, and a lack of data control. Companies can create consistency between the data sets of various departments by linking their general ledger to a single accounting system. Data requirements differ among departments, yet all parties benefit from pulling information within a consolidated, complete set of data. Leveraging its use of technology significantly impacts a tax department's ability to achieve a higher performance level and move beyond its role as a cost center. In their present form, tax controls and competencies rely heavily on individuals' skill sets. As a result, problems arise if staff turnover is high or available resources are mixed inappropriately. The automation of processes and consolidation of systems not only saves time and enables employees to contribute to more value-add activities, but it also leads to increased quality control.

A tax department loses a great deal of time as regulatory requirements tighten audit processes and increase the need for careful documentation. Detailed documentation ensures the presence of proper financial controls. Problematic audits arise when a company's various branches and entities lack integration or use several different accounting and tax systems. Misaligned data leads to wasted steps, thus wasted time and money, and generates risk of reporting errors, penalties, interest, and additional fees. Automation puts controls in place to monitor and better control data and information flow. This control improves audit trails and reduces risk levels.

Departments can effectively manage data and mitigate tax risk through carefully crafted tax processes involving technology and automation. The organization potentially faces high implementation costs; nevertheless, the resulting benefit of risk reduction offsets those costs. The department can detect errors more rapidly, avoid penalties and interest, and improve return on technological investment. Though technology in itself does not guarantee improvement, it assists in the betterment of the organization when applied appropriately. When mapped out, return completion and compliance are simply a function of processes. Consequently, the tax department can direct technological improvements appropriately by gaining a complete understanding of compliance processes and determining specific points of inefficiency. The department then can reduce and eliminate these inefficiencies.

A focused and concerted effort to improve processes around the areas of tax planning, tax risk management, and audit defense results in reducing both cost and time required to perform basic activity. At the same time, process improvement advances quality and reduces risk. It enables employees to contribute to additional high value-adding activities. Furthermore, as the tax function expands capability and increases productivity through internal process enhancements, improved technological resources, the department will develop a stronger and more collaborative relationship with other sectors of the corporation. As this indicates, the tax department contains potential to expand beyond its role as a cost center and instead become a profit center. Operations management techniques and strategies hold potential to transform the tax department into an efficient and effective revenue-generating contributor to the organization.

3. The future state of the tax department

Legislative developments, such as Sarbanes-Oxley, force the tax function to expand capacity beyond traditional core areas of compliance. In addition, since 2003, external forces such as new accounting standards and increased regulatory scrutiny require new capabilities among tax teams. In response to these dynamics, the tax department is undergoing a metamorphosis. Tax departments are recasting business processes, building stronger more heterogeneous leadership teams, and investing in new technology. The tax function is moving toward a more active role in supporting major transactions and operating decisions, contributing to internal controls and risk management initiatives, and collaborating with others on finance and accounting matters that are not directly linked to tax.

The tax department depends on management investment to avoid tax-related errors in financial statements and to contribute to tax planning activities. Additionally, investment contributes to building managerial and technical skills among senior tax executives. These skills are critical to the future success of the team. Consequently, management must be willing to expand investment if it would like to see the tax department move forward.

The department can immerse itself in high value-adding strategy and planning activity as it moves forward. Such immersion must be layered on top of the tax function's core compliance and tax planning activities. Lowering the effective tax rate through strategic tax planning could serve as a channel for tax to play a comprehensive strategic role. Other potential contributions include facilitating greater efficiency and standardization of processes, recording and analyzing data, and reporting. As a result, tax involvement early on in strategic business operations processes is worthwhile because the initial framing of business decisions will not only affect the overall success of the business but also affect their future core tax activities. For example, downstream activities, such as tax compliance, accounting, and record keeping, would be more efficient, less risky, and less time consuming for the tax department and for the company at large.

Funding, automation, and technology require additional input and tools to achieve optimal results in regard to tax efficiency. Organizational responsibilities and related tax resources must be aligned with the vision and objectives of the business to create exceptional results. A framework can provide several dimensions of analysis, including value drivers, processes, and enablers that help optimize tax decision making. In global organizations, management must examine decisions in light of the tax value drivers that align to broader corporate goals and objectives.

4. Key tax processes and tax process tools

In this section we introduce some of the key concepts and mathematical tools that can be effectively utilized in facilitating the implementation of the improvements discussed above.

Quantitative Tools:

1. Scheduling –Typically, each resource unit is scheduled for operation only a designated portion of total time (e.g., eight hours per day, five days per week). The amount of time a resource is scheduled for operation is called the **scheduled availability of the resource**. Scheduled availability of various resources in a process may differ. For example, in a manufacturing setting, some areas within a plant operate only one shift per day (8 hours) while others operate two (16 hours). The same time availability analysis could apply to the tax department due to the use of co-sourcing and outside consultants. Moreover, the choice of one day as the time period of measurement assumes that availability patterns repeat on a daily basis. More complicated patterns are possible as well. Some resource pools, for example, may be available only two days a week, with the pattern repeating every week. In that case, measure scheduled availability as the number of hours per week.

Taking into account the load batching and the scheduled availability of a given resource provides a general expression for the theoretical capacity of a resource unit:

Theoretical capacity of a resource unit = $(1/Tp)$ X Load Batch X Scheduled Availability (1)

The Theoretical Capacity of resource pool p (Rp) is given by the number of resources in resource pool p (Cp) times the theoretical capacity of a resource unit in a pool. This expression translates to:

$$Rp = (Cp/Tp) \text{ X Load Batch X Scheduled Availability} \qquad (2)$$

For example, consider a resource pool containing two tax managers, where each manager can handle a maximum of 10 tasks simultaneously. On average, the time to research and resolve each task is 15 minutes. Finally, assume that the two managers are scheduled to work 7.5 hours per day (450 minutes per day). Identify the following parameters to calculate the theoretical capacity of the pool:

$$Cp = \text{Number of resources} = 2 \text{ tax managers} \qquad (3)$$

$$Tp = \text{Unit Load} = 15 \text{ minutes} \qquad (4)$$

$$\text{Load batch} = 10 \text{ tasks to be completed} \qquad (5)$$

$$\text{Scheduled availability} = 450 \text{ minutes per day} \qquad (6)$$

Calculate the theoretical capacity of the two tax managers' pool using the Formula (2) above:

$$Rp = (2/15) \text{ X } 10 \text{ X } 450 = 600 \text{ tasks can be addressed per day by the group of two.} \qquad (7)$$

2.   Critical Path Analysis — The critical path method (CPM) is an algorithm for scheduling activities within a project. CPM determines the fastest and most efficient execution of that project. The algorithm, originally developed by DuPont and Remington Rand Corporation in the 1950s, is an essential project management technique. The CPM algorithm estimates the time necessary to complete each part of the project utilizing the information presented in the work breakdown structure and the precedence relationship and time estimates. All paths within a project need to be finished before the project can be considered complete. The **critical path**, also known as the bottleneck path or the binding constraint, is the path that takes the longest time to complete. This critical path dictates the project's duration. The activities making up a critical path are known as **critical activities.** Any delay in the execution of a critical activity results in delaying the entire project. The analyst also needs to understand how much slack, or flexibility, exists in scheduling noncritical activities. Slack is the estimate of the maximum amount of time that a noncritical activity can be delayed without affecting the entire project schedule. Therefore, analysts use a systematic algorithm to calculate the critical path and identify slack for each activity. A computerized systematic algorithm implements the CPM approach for large projects because they contain a sizeable number of paths and activities.

The algorithm for identifying the critical path and slack involves calculating the following four parameters for each activity:

1. Early Start Time (ES): The earliest time at which an activity can start, considering the beginning and ending times for each of the preceding activities.
2. Early Finish Time (EF): The sum of the early start time (ES) and the time required to complete the activity.
3. Late Start Time (LS): The latest time at which an activity can start, considering all of the precedence relationships, without delaying the completion time for the project.
4. Late Finish Time (LF): The sum of the late start time and the time required to complete the activity.

CPM requires calculations of the four parameters (ES, EF, LS, and LF) for each project activity. The implementation of the CPM algorithm begins by identifying ES and EF for all activities. The procedure starts with the first activity of a project, which has no predecessor. Once an analyst identifies ES and EF for the first activityhe repeats the same procedure for subsequent activity. . Relationships with previously completed activities must be taken into consideration beginning with the second activity. As a result, complete the calculations of ES and EF carefully so that none of the precedence relationships are ignored.

The essential technique for using CPM is to construct a model of the project that includes:

1. A list of all activities required to complete the project, typically categorized within a work breakdown structure.
2. The time (duration) that each activity will take to completion.
3. The requirements of each particular activity.

Using these values, CPM calculates the longest path of planned activities to the end of the project, and the earliest and latest that each activity can start and finish without making the project longer. This process determines which activities are "critical" (i.e. on the longest path) and which have "total float" (i.e. can be delayed without making the project longer). A critical path in project management is the sequence of project network activities that add up to the longest overall duration. This sequence determines the shortest time possible to complete the project. Any delay of an activity on the critical path directly impacts the planned project completion date (i.e. there is no float on the critical path). A project also can have several parallel, near critical paths. An additional parallel path through the network of total duration shorter than the critical path is called a sub-critical or non-critical path.

These results allow managers to prioritize activities for the effective management of project completion and to shorten the planned critical path of a project. Managers shorten the planned critical path by pruning critical path activities, "fast tracking" (i.e. performing more activities in parallel), and "crashing the critical path" (i.e. shortening the durations of critical path activities by adding resources). Managers can apply these concepts with success toward the optimization of the tax department.

Just as in manufacturing, principles exist to create flow in the tax function. The concept of a workflow cycle sets in place physical pathways and information flows and provides the timing of information flow down these pathways. As a result, tax personnel know where and when information should flow, and they can determine whether this flow is on time. Connecting processes with fixed pathways and binary signals creates this value stream flow. Essentially, individuals design a system for flow of data and information by flow paths with binary switches, similar to those in an electric circuit. Information and data are moved based on signals (completed/not-completed) along fixed pathways. Implementation of this technique proves successful, for instance, in regard to the timely and accurate review of tax provision. Certainly it takes time to design and implement a robust value stream flow, but once it is in place, the benefits dramatically outweigh this initial effort.

3. Constrained resource allocation--- Many project management techniques assume that a project team has the necessary resources to complete each activity within the specified function and budget. In reality, this might not be the case. Consequently, a number of supplementary techniques allocate resources to different project activities effectively. A resource breakdown structure and resource leveling are two commonly used resource management techniques. A **resource breakdown structure (RBS)** is a standardized list of personnel required to complete various activities in a project. This technique is often used in combination with the work breakdown structure. **Resource leveling** is an approach used to reduce the number of fluctuations in day-to-day resource requirements within an organization. This approach is especially useful when employees in an organization work on multiple projects simultaneously. Each project might have separate deadlines and a different set of resource requirements during different time periods. The number of employees typically remains constant. Therefore, the resource leveling approach is used to adjust the project schedule so that almost the same amount of personnel time is required every day to work on different projects.

Two problems arise in the deployment of scarce resources: the activity analysis problem and the optimal assignment problem

The Activity Analysis Problem. A company can employ n activities, A1, A2. ... ,An , using the available supply of m resources, R1, R2. ... , Rm (labor hours, computing capability, tax software availability, etc.). Let bi be the available supply of resource Ri. Let aij be the amount of resource Ri used in operating activity Aj at unit intensity. Let cj be the net value to the company of operating activity Aj at unit intensity. Choose the intensities with which the various activities are to be operated to maximize the value of the output to the company subject to the given resources.

Let xj be the intensity at which Aj is to be operated. The value of such an activity allocation is

$$\sum_{j=1}^{n} c_j x_j. \tag{8}$$

The amount of resource Ri used in this activity allocation must be no greater than the supply, bi; that is,

$$\sum_{j=1}^{n} a_{ij} x_j \leq b_i \qquad \text{for } i = 1, \ldots, m. \tag{9}$$

It is assumed that an activity at negative intensity cannot be operated; that is,

$$x_1 \geq 0, x_2 \geq 0, \ldots, x_n \geq 0. \tag{10}$$

Tax departments want to maximize (8) subject to (9) and (10). A company with scarce tax resources needs to address and optimize this standard maximum problem in its desire to advance its delivery of error free on time tax return deliverables and development of tax planning and tax strategy capabilities. A company also needs to address and optimize the assignment of these scarce resources.

The Optimal Assignment Problem. I persons are available for J jobs. The value of person i working 1 day at job j is aij , for i = 1, . . . , I, and

j = 1, . . . , J . Choose an assignment of persons to jobs to maximize the total value.

An assignment is a choice of numbers, xij , for i = 1, . . . , I, and j = 1, . . . , J, where xij represents the proportion of person i 's time that is to be spent on job j. Thus,

$$\sum_{j=1}^{J} x_{ij} \leq 1 \qquad \text{for } i = 1, \ldots, I \tag{11}$$

$$\sum_{i=1}^{I} x_{ij} \leq 1 \qquad \text{for } j = 1, \ldots, J \tag{12}$$

and

$$x_{ij} \geq 0 \qquad \text{for } i = 1, \ldots, I \text{ and } j = 1, \ldots, J. \tag{13}$$

Equation (11) reflects the fact that a person cannot spend more than 100% of his time working, (12) means that only one person is allowed on a job at a time, and (13) says that no one can work a negative amount of time on any job. Maximize the total value subject to (11), (12) and (13),

$$\sum_{i=1}^{I} \sum_{j=1}^{J} a_{ij} x_{ij}. \tag{14}$$

This optimal assignment calculation represents a standard maximum problem with m = I + J and n = IJ .

5. Terminology

The function to be maximized or minimized is called the objective function.

A vector, x for the standard maximum problem or y for the standard minimum problem, is said to be feasible if it satisfies the corresponding constraints.The set of feasible vectors is called the constraint set.

A linear programming problem is feasible if the constraint set is not empty. The problem is infeasible if the constraint set is empty.

A feasible maximum (resp. minimum) problem is unbounded if the objective function can assume arbitrarily large positive (resp. negative) values at feasible vectors; otherwise, the feasible maximum problem is considered bounded. Thus, three possibilities are available for a linear programming problem. The problem can be bounded feasible, unbounded feasible, or infeasible.

The value of a bounded feasible maximum (resp. minimum) problem is the maximum (resp. minimum) value of the objective function as the variables range over the constraint set. A feasible vector is optimal when the objective function achieves this value.

All Linear Programming Problems Can be Converted to Standard Form.

Maximizing or minimizing a linear function subject to linear constraints defines a linear programming problem. Analysts can convert all such problems into the form of a standard maximum problem by the following techniques.

A minimum problem can be changed to a maximum problem by multiplying the objective function by -1. Similarly, constraints of the form $\sum_{j=1}^{n} a_{ij}x_j \geq b_i$ can be changed into the form $\sum_{j=1}^{n}(-a_{ij})x_j \leq -b_i$. Two other problems arise.

(1) *Some constraints may be equalities.* An equality constraint $\sum_{j=1}^{n} a_{ij}x_j = b_i$ may be removed, by solving this constraint for some x_j for which $a_{ij} \neq 0$ and substituting this solution into the other constraints and into the objective function wherever x_j appears. This removes one constraint and one variable from the problem.

(2) *Some variable may not be restricted to be nonnegative.* An unrestricted variable, x_j, may be replaced by the difference of two nonnegative variables, $x_j = u_j - v_j$, where $u_j \geq 0$ and $v_j \geq 0$. This adds one variable and two nonnegativity constraints to the problem.

Deriving any theory for problems in standard form is applicable to general problems. Nevertheless, enlargement of the number of variables and constraints in (2) is undesirable from a computational point of view.

Tax departments can use these formulas and tools to manage time and activity scheduling and prioritize activities. The methods determine the optimal use of available resources.

6. Tax requirements based process design

Global businesses are complex and diverse organizations. Organizing work streams according to broad functional areas can help focus the process improvements necessary to achieve the targeted value drivers. Four primary enablers are critical to a high performing tax department: data and information, technology, well defined processes, and people. Quality data is the foundation for tax and finance decisions, while systems and technologies are required to capture, store, and maintain the integrity of data. Well established and robust processes facilitate the consistent use of those systems and technologies. The appropriate tax personnel strategically deploying the tools and resources provided to them expand the ability of the tax function to add value to the organization.

Tax Requirements Based Process Design (TRBPD) addresses the complexity of processes and data within the tax function. This tool provides the tax professional with a method to systematically diagnose the activities, events, and information flows of a work process. A work process might involve activities such as completing a net operating loss schedule, calculating the research & experimentation tax credit, or filing the federal or state tax returns for a corporation. The tool is used to analyze processes in order that all process stakeholders can improve their shared understanding, find improvement opportunities, solve process related problems, and optimize the process steps going forward. TRBPD is useful particularly when the tax professional desires further detail about the process and seeks to understand:

1. Who supplies the inputs into a process?
2. What constraints and requirements are placed on process inputs?
3. Who are the final customers? There can be both internal and external customers of the process?
4. What are the requirements of the customer?
5. Are customer requirements satisfied? If not, why not ?

TRBPD defines a process as a collection of requirements that must be satisfied in order to produce a work product and deliver a service that is ready for serving customer needs. The chief concern of TRBPD is final customer utility, and the primary tools used to help depict the TRBPD diagram are work swim lanes and the T-Square. T-squares illustrate each process step in terms of its key elements. They create tangible descriptions of a process step's inputs, outputs, and interval requirements. Users can define departmental accountability for each task in a process through work group swim lanes. Fully mapped out, T-squares and work group swim lanes enable process owners to visualize key events and working relationships. These key events include work products such as tax provision, and working relationships involve the interplay of differing tax types within the function. At each step, individuals can analyze process requirements (inputs), process work products (outputs), and determine who is accountable for carrying out a task.

The performance matrix is an additional tool used to build more effective tax processes. Performance matrices complement T-square maps and further detail process requirements. In addition, the performance matrix documents the key steps of a process, vital success factors, performance indicators, and process goals and standards. In addition, individuals successfully optimize constrained resource allocation problems in large tax departments with numerous complex problems through calculation procedures such as linear programming (LP). Together, the T-square map, the performance matrix, and linear programming serve as a foundation for designing and improving tax work processes.

7. Business Process Improvement (BPI) and the tax function

Organizations employ a number of formulas to improve business operations and get more of the colloquial "bang for their buck". Process mapping defines the existing system and serves as one of the operations management tools applied to clinically dissect the tax department's operations. Process mapping represents an essential first step to overlooked or superficially addressed opportunities for improvement. Effective programs invariably summarize an examination, analysis, and means of improvement for the business processes. A focused and concerted effort in improving processes within the tax department results in both reducing cost and time needed to perform the requisite tax function. This effort simultaneously improves quality and reduces risk. External players continually remind organizations of the need to understand their business processes. While focusing on business processes is not new, it has achieved some kind of vogue in the past several years.

Several programs, including reengineering, BPI, and Six Sigma, have emerged within that focus on improving business processes. Government mandates, such as Sarbanes-Oxley and PIPEDA, and certification organizations, such as ISO, have focused on process as well. Sometimes the external auditor, outside tax firm, or government agency identify control weaknesses. This external feedback serves as an excellent opportunity to understand what processes went wrong, how they went wrong, and how to reengineer improved, more robust processes. These efforts to improve processes and achieve greater efficiencies and higher quality deliverables will play a more central role as the tax function attempts to accomplish more high value add activity with fewer resources.

Individuals within the tax function recognize that understanding the details of their work holds value. They recognize a need to understand the mechanics of their process in order to complete their work more successfully. Conscientious tax team members dig into the details of their work, apply the tools available to them, and act to improve the process. Tax personnel likely possess a sufficient understanding of the work process if it is completed by a single individual, and they can make significant improvements to that process. On the other hand, a lack of exposure to proper tools and methods makes understanding more difficult when the work a person completes is a part of a larger process. Attempts to streamline a piece of the process in a vacuum, without consideration of the rest of the process, typically create additional problems rather than providing solutions.

Application of operations tools and methods to the tax department extends beyond U.S. regulatory compliance procedures. Activities such as transfer pricing utilize these methodologies on an international level. U.S. taxpayers frequently struggle with inefficient tax management practices in transfer pricing documentation. A taxpayer is subject to I.R.C. § 482 and the associated transfer pricing rules if the taxpayer is subject to U.S. income tax laws and has any business interaction with a commonly-controlled foreign affiliate. The purpose of §482 is "to place a controlled taxpayer on a tax parity with an uncontrolled taxpayer." The term "controlled" as used in § 482 includes any kind of control, direct or indirect, whether legally enforceable, and however exercisable or exercised. The reality of the control is the decisive component, not its form or the mode of its exercise. Under this broad approach, the vast majority of commonly controlled enterprises need to maintain up-to-date transfer pricing documentation.

Section 482 gives the Commissioner of the IRS authority to allocate income between two or more businesses "owned or controlled directly or indirectly by the same interests . . . if he determines that such . . . allocation is necessary in order to prevent evasion of taxes. . ." The Commissioner has broad discretion in making such allocations. These allocations will not be countermanded unless the taxpayer shows them to be unreasonable, arbitrary, or capricious. In addition, transactions between one controlled taxpayer and another are subject to special scrutiny to ascertain whether the common control is permissible in nature, or if the transaction is being used to reduce, avoid, or escape taxes.

U.S. law does not explicitly require taxpayers to maintain up-to-date transfer pricing documentation. Nevertheless, in the event the taxpayer is audited, heavy penalties will apply if the IRS determines that a transfer pricing adjustment is required. To avoid possible transfer pricing penalties, taxpayers must be able to demonstrate maintenance of contemporaneous transfer pricing documentation. IRC § 6662(e)(1)(b) and §6662(h) subject taxpayers to a potential penalty of 20 percent of the additional tax due under an assessment by the IRS. This penalty applies to taxpayers who substantially misstate the price of property or services exchanged in a controlled transaction to which § 482 applies. to. A "substantial misstatement" of price occurs when the taxpayer's stated transfer prices for goods or services are equal to 200 percent or more of the actual arm's length price, or when the prices are understated to at least 50 percent of the actual arm's length price, as determined by the IRS. A substantial misstatement of price also occurs if the net § 482 transfer price adjustment for the taxable year exceeds the lesser of $5,000,000 or 10 percent of the taxpayer's gross receipts. In addition, taxpayers who grossly overvalue or undervalue the price of goods sold in a controlled transaction are subject to a 40 percent penalty of the additional tax due under the assessment. A "gross valuation misstatement" of price occurs when the taxpayer's stated transfer prices for goods or services equal 400 percent or more of the actual arm's length price, or are understated to at least 25 percent of the actual arm's length price, as determined by the IRS. A gross valuation misstatement of price also will have occurred if the net § 482 transfer price adjustment for the taxable year exceeds the lesser of $20,000,000 or 20 percent of the taxpayer's gross receipts.

Many taxpayers may feel the temptation to neglect the task of keeping their transfer pricing documentation up to date because doing so requires a highly efficient international information management strategy. Recent accounting, financial, and economic pricing information must be identified and organized to truly develop contemporaneous transfer pricing documentation. The identification and organization of this information requires substantial effort and cost. Nevertheless, the threat of serious penalties typically provides sufficient grounds for continued compliance in this area. Furthermore, implementation of process improvements makes the procedure less tedious throughout the tax department as a whole.

8. Business process improvement in detail

Business process improvement, with a focus on the tax department, is the key to addressing the inefficiencies described above within the traditional tax function. At a very high level, BPI requires taking a broad view of information technology and business activity, as well as the relationships between them. Information technology is more than an automating or mechanizing force; it can fundamentally reshape the way business is done. By taking a process view to maximize effectiveness, business activities can be seen as more than a collection of individual or even functional tasks.

Information technology and BPI have a recursive relationship. Information technology capabilities should support business processes, and business processes should be developed in terms of the capabilities that the enabling technology can provide. This broadened view is often referred to as the recursive view of information technology, while BPI is considered the new industrial engineering.

BPI is rooted in information technology management, but the process can be defined most simply as a business initiative that has broad consequences in terms of satisfying the needs of customers and an organization's other constituents. BPI incorporates the skills of process measurement, analysis, and redesign The information systems group may need to play a behind-the-scenes advocacy role for the application of BPI to other departments. This advocacy likely involves convincing senior management of the power offered by information technology and process redesign. Specific business divisions lead BPI initiatives; information systems groups serve as partners in enabling these radical changes.

Applying BPI to the tax function involves several steps. First, the department must develop process objectives. BPI is driven by a business vision, which implies specific business objectives such as cost reduction, time reduction, and output quality improvement. These objectives might be enumerated in or implied by a company or department's strategic plan.

Secondly, the department needs to identify the processes to be improved. In the tax department's context, the identification of processes would be a primary responsibility of the vice president of tax and in conjunction with finance. Most organizations use the high-impact approach, which focuses on either the most important processes or those that conflict most with the business vision. A lesser number of organizations use the exhaustive approach, which attempts to identify all processes within an organization and prioritizes them in order of improvement urgency.

Third, the department avoids repeating prior mistakes and provides a baseline for future improvements by understanding and measuring existing processes. The initial activity of the tax department's working group's idea brings together project owners from the federal, international, state and local, and tax consulting and planning groups in order to cross-pollinate best practices and understand existing challenges and opportunities across the entire tax function.

Fourth, the department identifies information technology levers. An awareness of information technology capabilities can and should influence process design.

Fifth, the department can design and build a prototype of the new process based on these information technology capabilities. The actual design is not the end of the BPI process. Rather, the design is a prototype with successive iterations. The prototype aligns the BPI approach with quick delivery of results, as well as the involvement and satisfaction of the end-users and those relying on the tax information.

The next step is to define what the end goal should look like after key challenges and priorities for process improvement and reengineering are determined.

These goals may include improved year end and quarter end processes, integration of processes, optimal use of resources, automation of processes, understanding of critical tax paths, and quality control and assurance.

Tax accounting and compliance processes provide one of the greatest opportunities for process improvement. In tax accounting, the goal could be to improve the underlying accuracy of tax expense and deferred tax balances without allowing for any increase in the time to prepare the numbers. For compliance, the goal could be to improve the management of all tax compliance globally to ensure no mistakes are made.

Additionally, varied processes produce information for tax planning, tax accounting, and tax compliance in many businesses. Developing a single system that stores information from many different sources can be used to feed multiple applications. Integration of processes is essential to an efficient tax function.

Optimizing available resources also requires an understanding of how to go about deploying the optimal individuals for each task. Outsourcing certain tax consulting projects can enable the in-house tax function to focus on its core competencies and complete the required tax compliance in an efficient and effective manner, thus allowing for the function to take on additional strategic tax planning opportunities.

Tax departments traditionally use manual processes to accomplish required compliance-based tasks; nevertheless, automation of these processes could significantly benefit the tax function. Automation is of both improving the quality of tax deliverables and mitigating risk. Automation eliminates manual procedures, reduces the amount of human error, and allows tax personnel to focus on higher value-add activities. Automation also provides for higher quality data from the beginning, allows for a more controlled process, and limits manual intervention. Tax personnel need to define specific processes and consider what can be done to automate these. Automating defective processes does not prevent them from being defective; rather, automation is merely a tool to increase efficiency and effectiveness in the execution of designated tasks.

Tax and accounting personnel often face the "dumping" of miscellaneous expenses coded into wrong charge codes. The same applies with total hours worked in "general support" of one effort or another. Implementing integrated and automated processes relies on the quality of underlying source information being used for tax purposes. Process improvement often focuses on tax sensitivity, coding, and granularity of the base level data upon which the accounting information is based. As mentioned above, automation also lends itself to placing controls in place to monitor and better control data and information flow.

9. Implementation: how do we make it happen?

Tax personnel need to define process improvement projects on a case by case basis relative to the desired results. Successful process improvement projects require establishing business processes that contain several specific characteristics. Successful projects are efficient from a time and cost perspective in delivering the desired outcomes; possess built in controls that are able to be measured and tested; facilitate the organization's risk management objectives; are documented and can be easily communicated to the team and other functional teams, such as finance and operations; facilitate tax's ability to expand its utilization, adding value to finance and the executive suite in a decision making capacity; and provide monitoring and continuous evaluation of operational effectiveness.

In order for this to occur, an in-depth understanding must be gained regarding the current state of the department and the practices and procedures taking place. This understanding is a required starting point in order to determine which processes are working well and which are not. Mapping out processes diagrammatically provides a detailed illustration of process flows through the entire process. Where is the tax data originating? What manipulation is required? What is the quality of the data? What reconciliation and additional analytics need to be performed on the data? Which tax processes work well and which do not? What are the roles of the tax function? Are these roles too limited or too broad? Where are the inefficiencies? How should these inefficiencies be addressed?

After the current state has been determined, create a defined state that serves as a link between the current state and the future state.

This defined state forces the focus of the business improvement effort back on the department's end goals. Existing processes need to be reviewed and challenged and, if appropriate, a new process introduced. The new processes should include:

- Clearly defined and fully understood roles and responsibilities, both within tax and from supporting departments.
- Effective and clear project management oversight of operations processes.
- Properly addressed interdependencies between business units and corporate.
- Appropriate controls and checks to ensure data integrity.
- Overall managerial and project control points.
- Risk control and management processes to ensure that tax processes remain in line with objectives.

Lastly, successful process improvement requires a commitment from within tax and other departments to take ownership of their particular roles and responsibilities and to execute upon these roles.

A further operational method can be applied as this process takes place: gap analysis. Gap analysis is a tool that helps a company to compare its actual performance with its potential performance. Gap analysis is a formal study of what a business is doing currently and where it wants to go in the future. Two questions are at the core of gap analysis: "where are we?" and "where do we want to be?" If a company or organization is not making the best use of its current resources or is forgoing investment in capital or technology, it may be producing or performing at a level below its potential, or below its production possibilities frontier.

The goal of gap analysis is to identify the gap between the optimized integration of inputs and the current level of allocation. This identification helps provide the company with insight into areas of possible improvement. The gap analysis process determines, documents, and approves the variance between business requirements and current capabilities. Gap analysis naturally flows from benchmarking and other assessments. Once the general expectation of performance in the industry is understood, analysts can compare that expectation with the company's current level of performance. This comparison becomes the gap analysis. Individuals can be perform such analysis at the strategic or operational level of an organization.

Gap analysis can be conducted in different perspectives: organization (i.e. the tax department), business direction, business processes, and information technology. Once personnel define the new desired state, they should properly test any new systems before rolling out or going live. A test run with a focus on fixing any remaining issues is necessary on any new systems before their full implementation occurs. Running the new and old processes in parallel after implementation changes take place is a worthwhile effort, especially with tax accounting projects. Often the preferred solution is unlikely to be perfect, and a continuous process improvement process needs to be embraced. The complex environment of the tax function contributes to the need for the integration of this continuous improvement process. Furthermore, periodically monitoring and reviewing a new system distinguishes potential weaknesses and future improvement opportunities.

Finance and tax executives frequently state a clear desire to increase efficiencies and provide higher value contributions. In particular, management recognizes a need to develop a role for the tax function that focuses on more than just traditional compliance and tax planning. This role moves beyond the current, narrow scope to contribute to the strategic practices of the enterprise as a whole and encompass higher value adding activity. In particular, tax needs to step in early on to support major transactions. Supporting major transactions consists of working to optimize the transaction structure and timing with the downstream impact on tax positions, and to support operating decisions. Strengthening leadership and management as well as technical tax capabilities grant the tax team the ability to undertake these objectives in a successful manner.

Changes in tax law, corporate structure, and regulators' oversight of results drive the tax department to perform more effectively and efficiently. Strategic technology investments support sound decision making, and sound business process improvement drives higher quality information and lower risk.

Organizations achieve this increase in effectiveness and efficiency through four primary mechanisms:

1. Thoroughly understanding and optimizing tax processes from an operations viewpoint.
2. Investing in technology for tax.
3. Strengthening the leadership of tax.
4. Honing business processes of the tax function as they relate and interact with operations, finance, and technology within the enterprise.

Optimizing strategic value from within the corporate tax function requires senior tax personnel to acquire new, broader, and somewhat different skill sets. The best practice among leading companies addressing the issue of financial statement errors involves the tax function from the outset in major transactions and operating decisions. In doing so, the tax function surfaces the tax ramifications of those transactions early on and highlights any potentially hazardous tax issue that could result in significant financial statement restatements in the future. This type of contemporaneous activity serves as the hallmark of a strategically integrated tax department as it facilitates increased efficiency and effectiveness in broadening tax's role across the entire company.

10. Concluding remarks

The tax function can optimize its role within an organization by utilizing operations management tools and methods, as well as through the study and implementation of business process improvements. The application of these tools serves as a means of both enabling a team to complete compliance-related tasks with greater efficiency and integrating the corporate tax department with other functions of an organization.

Organizational structure historically segregates the tax department within the finance function. This segregation is often coupled with a lack of understanding among finance executives of the tax function's potential contribution to business operational and strategic issues. At the most basic level, integrating process improvements provides tax employees with additional time that can be spent on strategy and planning rather than compliance. Tax's highly specialized and fairly distinct domain focuses primarily on complying with complex tax requirements. Adhering to different calendars, based on the timetables of the IRS, further detaches the tax function from other departments. Furthermore, tax's processes and systems, accounting methods, and reporting standards differ from those used for financial and managerial reporting. Thus the tax department's unique demands and specialized knowledge isolate it from the rest of the finance function and the business at large.

Nevertheless, recent changes in tax law, financial reporting transparency, and corporate governance requirements drive a change in how the tax function, finance function, and broader management team work together in achieving a corporation's objectives on behalf of its stakeholders. The Sarbanes-Oxley Act, for example, contributes as a major driver of this change as it requires companies to document their business process controls. Increased standardization and automation of processes furthers this transformation. It reduces waste, increases efficiency, maintains continuous flow and allows for greater participation by tax in its role as a strategic partner. Additional transformational contributions include new delivery models such as shared services, outsourcing, and business process reengineering. The integration of operations management methods and practices, in particular through the automation and improvement of business processes, as well as the study and identification of specific points of improvement, leads a tax department to operate with decreased complexity, greater productivity, and as an essential contributor to overall business strategy.

11. References

An Introduction to Linear Programming, Steven J. Miller, March 31, 2007, Mathematics
    Department, Brown University, 151 Thayer Street, Providence, RI 02912
Cogliandro, J. (2007) Intelligent Innovation, *J.Ross Publishing*, Fort Lauderdale, FL.
Gibson, R. and Skarzynski, P. (2008), Innovation to the Core, *Harvard Business Publishing*,
    Boston, MA.
Holtzman, Yair, Innovation in Research and Development: Tool of Strategic Growt, *Journal
    of Management Development* p. 1037-1052, JMD Vol. 27, Number 10, 2008
Mandelbaum, A. (1996), "Getting the most out of your product development process",
    Harvard Business Review, March-April.
The Journal of Product Innovation Management (2002) Implementing a strategy-driven
    performance measurement system for an applied research group, Christoph H.
    Loch, U.A, Staffan Tapper, INSEAD, Fontainebleau, France & Witts Graduate
    School of Business, Johannesburg, South Africa

Part 3

Mathematical Methods
for Evaluating Operations

7

Some Remarks About Negative Efficiencies in DEA Models

Eliane Gonçalves Gomes[1], João Carlos C. B. Soares de Mello[2],
Lidia Angulo Meza[2], Juliana Quintanilha da Silveira[2],
Luiz Biondi Neto[3] and Urbano Gomes Pinto de Abreu[4]

[1]*Brazilian Agricultural Research Corporation - Embrapa*
[2]*Fluminense Federal University*
[3]*Rio de Janeiro State University*
[4]*Embrapa Pantanal*
Brazil

1. Introduction

Data Envelopment Analysis appeared in 1978 when the first model, known as CCR was proposed by Charnes et al. (1978). This model calculates the efficiency of productive units, known as DMUs - Decision Making Units, by comparing the use of resources (inputs) and the production (outputs) obtained. This model considers Constant Returns to Scale (CRS), i.e., an increase in resources generates a proportional increment to products. This proportion is constant for all DMUs. An important issue regarding the original CCR model is that all data and variables must be non-negative.

The BCC model proposed by Banker et al. (1984) introduced the hypothesis of variable returns to scale to the DEA models. In this model there is no proportionality between increments in resources and the correspondent increments in products. Among its other characteristics, this model allows the use of negative variables due to some invariability properties in the data translation procedure (Pastor, 1996, Thrall, 1996, Iqbal Ali & Seiford, 1990). In addition, Sharp et al. (2007), Duzakin and Duzakin (2007), Portela et al. (2004) and Sueyoshi (2004) presented examples of DEA models that deal with negative outputs or inputs, including some sophisticated theoretical formulations.

The efficiency achieved by the DEA models was always taken as non negative. In the specific case of classic input orientated models (radials, CCR or BCC), the immediate result of the objective function is the efficiency, measured in the interval [0,1]. This is an immediate consequence of the formulation of the multipliers model, because a set of restrictions establishes that the weights or multipliers are calculated in such a way as to maximize the efficiency of the DMU under analysis. However, a second group of restrictions determines that those multipliers, when used to evaluate other DMUs, cannot generate efficiencies higher than 1. So given that all of the multipliers are not negative, and that the variable data are not negative either, the weighted sum that calculates the efficiency is measured in the interval [0,1].

In the case of the DEA-BCC model, the multipliers model was originally obtained through the dual of its envelope formulation. This model holds an equality restriction such that it finds a convex frontier. The multipliers model in the dual envelope method has a free variable. This means that we cannot guarantee the non negativity of the efficiencies when the multipliers of a DMU are used to evaluate another DMU. When working with the classic DEA models this is not a problem, but negative efficiencies can be generated in typical situations of Cross Evaluation (Sexton et al., 1986, Doyle & Green, 1994) or in the calculation of some non-radial efficiencies (Lins et al., 2004, Quariguasi Frota Neto & Angulo-Meza, 2007).

In this chapter it will be shown that the input orientated DEA BCC model can generate negative efficiencies, which are usually hidden in the model. Using a two-dimensional example (one input and one output), the condition for the possible occurrence of negative efficiencies will be shown. Furthermore, we will show that a small intuitive change in the BCC multipliers model fixes this situation. However, we will show that this modification generates a major change in the dual envelope model, producing alterations to the frontier. We give an interpretation for this new dual model using a non-observed DMU.

Numerical examples will be presented. The first example has two variables, one input and one output. It will be geometrically shown that the modified model causes a change in the efficient frontier. These changes will be interpreted by the introduction of a non-observed DMU. In this first example, the DMU causing negative efficiency in other DMUs has a reduction of its efficiency in the modified model. This reduction does not occur in the second example (two inputs and one output). This happens due to the optimum multipliers set multiplicity.

We will show that the negative efficiencies mentioned in this chapter apply neither to the CCR model nor to the output orientated BCC model. The latter has a restriction in the multipliers model that is of the "greater than" or "equal to" type, ensuring the non negativity of the efficiency measures.

Aside from simple numerical examples, we will also use the proposed theoretical approach to assess the efficiency of cattle breeders in some Brazilian municipalities

2. DEA models - general aspects

DEA uses mathematical programming problems to estimate a piecewise linear efficient frontier. DEA can deal with multiple inputs and outputs to calculate the efficiency of the firms, or production units, or DMUs. DEA optimizes each individual observation in order to estimate an efficient frontier (piecewise linear), composed of the units with the best practices within the evaluation sample (Pareto-Koopmans efficient units). These firms are references or benchmarks for the inefficient ones.

There are two classic DEA models. The CCR model (also known as CRS or constant returns to scale), which deals with constant returns to scale (Charnes et al., 1978) and assumes proportionality between inputs and outputs. The BCC model (or VRS), due to Banker et al. (1984), assumes variable returns to scale, i.e. replaces the axiom of proportionality by the axiom of convexity (Lins & Angulo-Meza, 2000). Traditionally, there are two orientations to these models: input oriented, if we want to minimize the resources available, without changing the level of production; oriented to outputs, when the goal is to increase the production, without changing the amount of inputs used. See Cooper et al. (2004) for details. A usual DEA assumption is that resources and products are subject to physical measurement. However, this is not strictly necessary and proxies can be used (Souza, 2006).

The CCR model constructs a non-parametric piecewise linear frontier involving the data. As mentioned before, it assumes constant returns to scale, that is, any change in inputs produces proportional variation in the outputs. This model determines the efficiency by dividing the weighted sum of the outputs (virtual output) by the weighted sum of the inputs (virtual input), generalizing thus the definition of Farrell (1957). The model allows each DMU to choose the weights for each variable (input or output) in the way that is more benevolent. These weights when applied to other DMUs cannot produce a ratio greater than 1. These conditions are formalized in (1), where each DMU k, $k = 1...n$, is a production unit that uses r inputs x_{ik}, $i = 1...r$, to produce s outputs y_{jk}, $j = 1...s$; x_{i0} and y_{j0} are the inputs and outputs of DMU 0; u_j and v_i are the weights calculated by the model for inputs and outputs, respectively.

$$\text{Max } \frac{\sum_{j=1}^{s} u_j y_{jo}}{\sum_{i=1}^{r} v_i x_{io}}$$

subject to

$$\frac{\sum_{j=1}^{s} u_j y_{jk}}{\sum_{i=1}^{r} v_i x_{ik}} \leq 1, \quad k = 1...n$$

$$v_i, u_j \geq 0, \quad i = 1...r, \ j = 1...s$$

(1)

The fractional programming problem (1), which must be solved for each DMU, can be transformed into a Linear Programming Problem (LPP). To do so, we may impose that the denominator of the objective function should be equal to a constant, usually unity. This linear formulation of the CCR model is presented in (2) and is called Multipliers Model with input orientation.

$$\text{Max } \sum_{j=1}^{s} u_j y_{jo}$$

subject to

$$\sum_{i=1}^{r} v_i x_{i0} = 1$$

$$\sum_{j=1}^{s} u_j y_{jk} - \sum_{i=1}^{r} v_i x_{ik} \leq 0, \ k = 1...n$$

$$v_i, u_j \geq 0, \quad i = 1...r, \ j = 1...s$$

(2)

There are two equivalent, dual, formulations for DEA. Simply put, we can say that one of the formulations, the Multipliers model, deals with the ratio of weighted sums of products and resources, with the weighting chosen to be more favourable to each DMU, subject to certain conditions. The other formulation, the Envelope model, defines a feasible region of production and works with projections of each DMU in this frontier.

In (3) we present the DEA CCR Envelope model, input oriented, where θ_0 is the efficiency of the DMU under analysis, λ_k represents the contribution of DMU k to obtain the target for DMU 0.

$$\text{Min} \quad \theta_0$$
$$\text{subject to}$$
$$\theta_0 x_{i0} - \sum_{k=1}^{n} x_{ik}\lambda_k \geq 0, \quad i = 1...r$$
$$-y_{j0} + \sum_{k=1}^{n} y_{jk}\lambda_k \geq 0, \quad j = 1...s \tag{3}$$
$$\lambda_k \geq 0, \quad k = 1...n$$

The DEA-BCC model was proposed for dealing with situations whereby the proportionality between inputs and outputs is not constant along the efficient frontier, thereby forming a variable returns to scale (VRS) frontier. The BCC model was originally obtained by adding a convexity restriction to the formulation of the CCR envelope model (3). This new restriction is $\sum_{k=1}^{n} \lambda_k = 1$. The frontier is piecewise linear and takes the different production scales into account: increasing, constant and decreasing returns to scale. The input orientated BCC multipliers model - the dual of the Envelope Model - is presented in (4).

$$\text{Max} \quad \sum_{j=1}^{s} u_{j0} y_{j0} + u_*$$
$$\text{subject to}$$
$$\sum_{i=1}^{r} v_{i0} x_{i0} = 1$$
$$\sum_{j=1}^{s} u_j y_{jk} + u_* - \sum_{i=1}^{r} v_i x_{ik} \leq 0, \quad k = 1...n \tag{4}$$
$$v_i, u_j \geq 0, \quad u_* \in \Re$$

In these models, the variable u_* indicates if the observed DMU is found in the area of increasing, constant or decreasing returns to scale. This variable also represents the independent term in the supporting hyper-plane equation and can assume any real value as shown in Fig. 1. As was previously mentioned, this is the dual variable corresponding to the envelope model equality restriction.

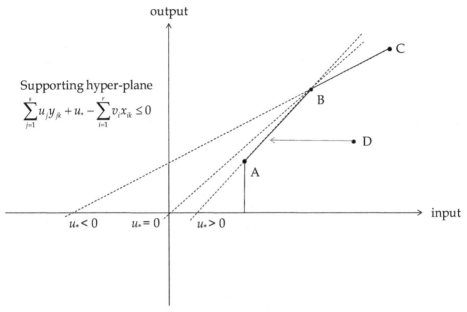

Fig. 1. Supporting hyper-plane to the input orientated BCC model.

Fig. 1 shows the supporting hyper-plane equations for different values of the variable u_*. When this term is strictly positive the DMU works in the increasing returns to scale part of the efficient frontier. When it is equivalent to zero it has constant returns, and when is strictly negative it is in the decreasing returns to scale part of the efficient frontier.

In order to model and to interpret correctly the DEA results it is necessary to know their models properties. Two of the most important are (Gomes et al., 2009):

- In any DEA model, the DMU that has the best value of the ratio $(output\ j)/(input\ i)$ will always be efficient. This property requires the existence of a causal relationship between each output and each input. Ignoring this relationship can lead to meaningless results.
- The CCR model, which in its fractional form is a zero degree homogeneous function. Its main property is the proportionality between inputs and outputs at the frontier. As a consequence, an increase (decrease) in the amount of inputs will cause a proportional increase (decrease) in the value of their outputs.

3. Negative efficiencies in the DEA BCC model

In model (4), it should be noted that when the multipliers of the DMU 0 are used to evaluate other DMUs, the expression $\sum_{j=1}^{s} u_{j0}y_{j0} + u_{*0}$ may be negative when u_{*0} is sufficiently negative.

This means that the efficiency of DMU j, when evaluated with the multipliers of the DMU 0, may be negative. This may occur when the DMU 0 is in the decreasing returns to scale portion of the frontier. Negative efficiencies were detected, but not studied, by Soares de Mello et al. (2002).

In the case of just one *output*, $r = 1$, we are able to derive a condition to determine when negative efficiencies will appear in the restrictions for a specific DMU j. This happens when $u_{10}y_{1j} + u_{*0} < 0$, that is, $y_{1j} < -u_{*0}/u_{10}$.

Table 1 shows both data and results of a numerical example with one input and one output, illustrating the situation previously described. It is important to highlight that multiple optimum multipliers exist for efficient DMUs (Rosen et al., 1998, Nacif et al., 2009, Soares de Mello et al., 2002) and we here used the first multipliers found by the software SIAD (Angulo-Meza et al., 2005). Note that in Table 1 the u_{*0} is a negative value for DMU A.

DMU	Input	Output	Multipliers			Efficiency
			v	u	u_*	
A	4	10	0.250	0.375	-2.750	1.000
B	1	5	1.000	0.000	1.000	1.000
C	2	7	0.500	0.000	0.500	0.500
D	1	8	1.000	0.125	0.000	1.000
E	6	6	0.167	0.000	0.167	0.167

Table 1. Numerical example – one input and one output.

Table 2 presents the DMUs' efficiencies calculated using the multipliers of the others, in an approach similar to the Cross Evaluation (Sexton et al., 1986, Doyle & Green, 1994). In this table the values in a column j are the efficiencies of the DMU j when evaluated using the DMU of the respective row.

	A	B	C	D	E
A	1.000	-3.500	-0.250	1.000	-0.333
B	0.250	1.000	0.500	1.000	0.167
C	0.250	1.000	0.500	1.000	0.167
D	0.313	0.625	0.438	1.000	0.125
E	0.250	1.000	0.500	1.000	0.167

Table 2. Cross efficiencies for the numerical example presented in Table 1.

In Table 2, we can observed that when using the output multiplier of the DMU A $u_A = 0.375$, and the independent term, $u_{*A}=-2.750$, three negative efficiencies are obtained. This happens for DMUs B, C and E. The outputs of these DMUs are less than 7.333 (result of $-u_{*A}/u_A$). These results have no interpretation in the classic theory of efficiency. Furthermore, these implicit negative efficiencies may be the main reason for the use of Cross Evaluation only with constant returns to scale (CRS) DEA models

4. Proposed DEA BCC model with non-negativity constraint

The problem of negative efficiencies can be easily solved by imposing an additional constraint for each DMU, as shown in model (5). This model is called the Modified Input

Orientated DEA BCC Model. It should be mentioned that a similar approach for specific cases was proposed by Wu et al. (2009) and Angulo-Meza et al. (2004), without meaningful interpretation.

$$\text{Max } \frac{\sum\limits_{j=1}^{s} u_j y_{jo} + u_*}{\sum\limits_{i=1}^{r} v_i x_{io}}$$

subject to

$$0 \le \frac{\sum\limits_{j=1}^{s} u_j y_{jk} + u_*}{\sum\limits_{i=1}^{r} v_i x_{ik}} \le 1, \quad k = 1...n$$

$$v_i, u_j \ge 0, u_* \in \Re, \quad i = 1...r, \ j = 1...s$$

(5)

The linear form of model (5) is presented in (6). In this model, each restriction $\sum\limits_{j=1}^{s} u_j y_{jk} + u_* \ge 0$ would be called a non-negativity restriction.

$$\text{Max } \sum\limits_{j=1}^{s} u_j y_{jo} + u_*$$

subject to

$$\sum\limits_{i=1}^{r} v_i x_{io} = 1$$

$$\sum\limits_{j=1}^{s} u_j y_{jk} + u_* - \sum\limits_{i=1}^{r} v_i x_{ik} \le 0, \quad \forall k$$

$$\sum\limits_{j=1}^{s} u_j y_{jk} + u_* \ge 0, \quad \forall k$$

$$v_i, u_j \ge 0, u_* \in \Re, \quad \forall i, j$$

(6)

Table 3 shows the results of model (6) using the data in Table 1. The results show that the Modified Input Oriented BCC Model has changed the efficiency of DMU A. According to this new model, DMU A has become inefficient, because it was the only one that generated negative efficiencies when used to evaluate other DMUs. In fact, the term $-u_{*A}/u_A$ of DMU A is equal to 5.00, being the lowest output value in the set of DMUs.

Once again we used the multipliers obtained with model (6) to calculate the Cross Evaluation Matrix for all DMUs. The results are presented in Table 4. Observe that all of the evaluations carried out by all DMUs generate efficiencies between 0 and 1. It is interesting to note that DMU B, when evaluated by DMU A has an efficiency of 0, i.e. the non-negative restriction is an active restriction. In classic DEA models the null efficiency can only appear when all of the outputs are null.

DMU	Input	Output	Multipliers			Efficiency
			v	u	u_*	
A	4	10	0.250	0.083	-0.417	0.417
B	1	5	1.000	0.000	1.000	1.000
C	2	7	0.500	0.000	0.500	0.500
D	1	8	1.000	0.000	1.000	1.000
E	6	6	0.167	0.000	0.167	0.167

Table 3. Numerical example for the modified DEA-BCC model.

	A	B	C	D	E
A	0.417	0.000	0.333	1.000	0.056
B	0.250	1.000	0.500	1.000	0.167
C	0.250	1.000	0.500	1.000	0.167
D	0.250	1.000	0.500	1.000	0.167
E	0.250	1.000	0.500	1.000	0.167

Table 4. Cross Efficiencies for the modified DEA BCC model.

5. Interpretation of the additional restriction

At this point it is important to discuss the effects on the efficient frontier caused by the new restriction. In Thanassoulis & Allen (1988) it was shown that multipliers restrictions can be replaced by one or more unobserved or artificial DMUs, i.e. DMUs that do not exist in the original data set. As the restrictions of non-negativity, one for each DMU, are in fact multipliers restrictions, they can be replaced by unobserved DMUs. So, the efficient frontier in the model with non-negativity constraints may also depend on inefficient DMUs. Specifically, the inefficient DMUs that may change the frontier are those with, at least, one negative cross efficiency in the classic model.

Analysing the results of the numerical example, we observe that DMU A was the only one whose efficiency was altered by model (6). As a result, the modified model frontier may be obtained including a non-observed DMU. The output of this non-observed DMU is the same output of DMU A. The input of the non-observed DMU is obtained by multiplying DMU A actual input by DMU efficiency in the modified model. For our numerical example the non-observed has an output of 10 and an input of 1.6667.

Fig. 2 shows both DEA BCC and modified DEA BCC frontiers. In this Figure the bold line represents the two models common frontier, the dashed line belongs only to the DEA BCC model and the dotted line represents the frontiers of the modified DEA BCC model. We clearly observe that the new frontier is dislocated from the original DEA BCC frontier.

The geometric representation here above is valid only for the one input and output case. The multidimensional case can be interpreted only using the dual formulation of the modified BCC model. This formulation is presented in (7).

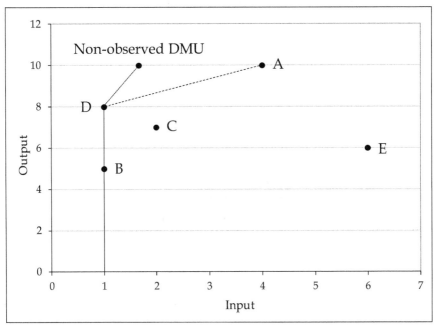

Fig. 2. DEA BCC frontier and the modified DEA BCC frontier.

$$\text{Min} \quad \theta_0$$
$$\text{subject to}$$
$$\sum_{k=1}^{n} y_{jk}\lambda_k - \sum_{k=1}^{n} y_{jk}\lambda'_k \geq y_{j0}, \quad \forall j$$
$$\theta_0 x_{i0} \geq \sum_{k=1}^{n} x_{ik}\lambda_k, \quad \forall i \qquad (7)$$
$$\sum_{k=1}^{n} \lambda_k - \sum_{k=1}^{n} \lambda'_k = 1, \quad \forall k$$
$$\lambda_k, \lambda'_k \geq 0, \quad \forall k$$

The additional restrictions of model (6) - one for each DMU, generate the same number of additional decision variables in the dual model (7). These variables are called λ'. It should be noted that when the sum of all λ' is not null, then the sum of λ won't be unitary. Therefore, in this case, there won't be any guarantee of convexity in the modified model.

From the Complementary Slack Theorem, we know that λ' times the corresponding restriction slack must be null. Consequently, λ' can only be other than zero, if the slack in the corresponding additional restriction of model (6) is null. This happens when the corresponding additional restriction is active. This means that the DMU corresponding to the active restriction would have negative efficiency using DMU 0 multipliers.

In the previous numerical example, only DMU A had its efficiency changed due to the additional restrictions, and this change was caused by the non-negativity restriction relative to DMU B. So, all λ' for DMU A are null, except λ'_B.

6. Three-dimensional numerical example

To illustrate the modified BCC model in situations with more than two variables, we will present a new numerical example. Table 5 shows data for the numerical example with seven DMUs, two inputs and one output. The variables' multipliers and efficiencies are depicted in the same table. As previously done, the multipliers shown are the first found by the SIAD software.

| DMU | Input1 | Input2 | Output | Multipliers | | | | Efficiency |
				v_1	v_2	u	$u*$	
A	0.489	0.637	0.607	0.300	1.340	1.627	0.012	1.000
B	1.000	1.000	1.000	1.000	0.000	1.300	-0.300	1.000
C	0.019	0.190	0.010	30.702	2.193	0.000	1.000	1.000
D	0.032	0.008	0.005	30.702	2.193	0.000	1.000	1.000
E	0.096	0.052	0.032	0.000	19.231	19.173	0.058	0.672
F	0.053	0.035	0.007	17.754	1.687	15.240	0.505	0.612
G	0.898	0.164	0.115	0.000	6.098	6.079	0.018	0.717

Table 5. Data and results for the three-dimensional numerical example.

As we can observe, the $u*$ of DMU B is negative. When the multiplier of the sole output ($u = 1.30$) and the independent term ($u* = -0.300$) from DMU B are used to evaluate the other DMUs, negative efficiencies will appear for DMUs that have an output value inferior to 0.231. The DMUs that have output inferior to this value are C, D, E, F and G, whose efficiencies will be negative. The Cross Evaluation Matrix (Table 6) illustrates these comments.

The results of the modified BCC model are depicted in Table 7. Note that the modified model didn't change the efficiency of any DMU. This happened because DMU B, which generated negative efficiencies when evaluating the others, is also efficient in the modified model. This is due to the multiplicity of the optimum multipliers set previously mentioned.

	A	B	C	D	E	F	G
A	1.000	1.000	0.109	1.000	0.653	0.375	0.408
B	1.000	1.000	-15.119	-9.180	-2.694	-5.493	-0.168
C	0.061	0.030	1.000	1.000	0.327	0.587	0.036
D	0.061	0.030	1.000	1.000	0.327	0.587	0.036
E	0.955	1.000	0.068	1.000	0.672	0.286	0.717
F	1.000	0.810	1.000	1.000	0.554	0.612	0.139
G	0.955	1.000	0.068	1.000	0.672	0.286	0.717

Table 6. Cross evaluation matrix for the three-dimensional numerical example.

DMU	Input1	Input2	Output	Multipliers				Efficiency
				v_1	v_2	u	u_*	
A	0.489	0.637	0.607	0.300	1.340	1.627	0.012	1.000
B	1.000	1.000	1.000	0.183	0.817	0.993	0.007	1.000
C	0.019	0.190	0.010	26.989	2.564	23.168	0.768	1.000
D	0.032	0.008	0.005	30.525	2.900	26.203	0.869	1.000
E	0.096	0.052	0.032	0.000	19.231	19.173	0.058	0.672
F	0.053	0.035	0.007	17.754	1.687	15.240	0.505	0.612
G	0.898	0.164	0.115	0.000	6.098	6.079	0.018	0.717

Table 7. Data and results for the numerical example using modified BCC model.

The Cross Evaluation Matrix for the modified DEA BCC model is shown in Table 8. Note that there are no negative cross efficiencies in this table.

	A	B	C	D	E	F	G
A	1.000	1.000	0.109	1.000	0.653	0.375	0.408
B	1.000	1.000	0.109	1.000	0.653	0.375	0.408
C	1.000	0.810	1.000	1.000	0.554	0.612	0.139
D	1.000	0.810	1.000	1.000	0.554	0.612	0.139
E	0.955	1.000	0.068	1.000	0.672	0.286	0.717
F	1.000	0.810	1.000	1.000	0.554	0.612	0.139
G	0.955	1.000	0.068	1.000	0.672	0.286	0.717

Table 8. Cross evaluation matrix for the Modified DEA BCC model from the three-dimensional example.

7. Case study

We will use the modified DEA BCC model to evaluate the efficiency of some livestock systems.

The central structure in the beef cattle production chain is the biological system of beef production, including the various stages of creation (cow-calf production, stocker production, feedlot beef production) and their combinations, around which the producers are grouped. In Brazil, the cow-calf beef cattle phase occurs predominantly in an extensive continuous grazing, with native and/or cultivated pastures, encompassing: calves (until weaning or even one year old), cows, heifers and bulls. The cow-calf phase is the lower profitability activity and the one of major risk. However, it supports the entire structure of the beef production chain.

This case study, by using the Modified DEA BCC models here proposed, seeks to assess the comparative performance of extensive livestock modal production systems in its cow-calf phase, in some municipalities of Brazil. The objective is to measure the performance of the cattle farmer's decision regarding the composition of the production system, which has a direct impact on the expenditures and on the income generated. A study carried out with the same dataset and based on DEA BCC model with weights restrictions can be found in Gomes et al. (2011).

7.1 Data source

Primary data were collected through the panel system, which allows the definition of representative farms, as proposed by Plaxico and Tweeten (1963).

Despite the difficulty of characterizing a single property and a production system that is representative of the locality under study (here the city/municipality), this method looks through the experience of the participating farmers to characterize the property that is the most commonly found in the region. In some cases, the impossibility of determining this typology imposes the specification of more than one representative property or production system.

The panel is a less costly procedure of obtaining information than the census or the sampling of farms. Another advantage is that it provides greater flexibility and versatility in data updating, without affecting their quality. The technique consists in a meeting with a group of one or more researchers, one technician and eight regional farmers, on average (it can range from five to ten). Meetings are scheduled in advance, with the support of rural unions and regional contacts. The subjects and numbers, determined previously in interviews with local technicians, are discussed with the farmers. At the end of that debate, one can say that any characterization of the typical farm in the region has the consent of the farmers. Thus, productivity rates, establishment costs, fixed and variable costs, i.e., all the numbers resulting from the panel, tend to be fairly close to the regional reality.

It is noteworthy that the rates and the costs reported by each participant are not related to their properties, but with a single farm, declared at the beginning of the panel as the one that best represents the scale of operation and the production system of most of the local properties.

This study evaluated 21 beef cattle modal production systems that performed only the cow-calf phase, in seven states of Brazil. The data, derived from the indicators of the project developed by the Centro de Estudos Avançados em Economia Aplicada and the Confederação da Agricultura e Pecuária do Brasil, were collected in municipalities of these seven states: Mato Grosso do Sul - MS (eight), Goiás - GO (four), Rio Grande do Sul - RS (one), Minas Gerais - MG (four), Tocantins - TO (two), São Paulo - SP (one) and Bahia - BA (one). Panels with the farmers, with the support of the local rural technical assistance, were performed to collect the data, according to the methodology described in Centro de Estudos Avançados em Economia Aplicada (2010).

7.2 Modelling

7.2.1 DMUs

The objective of the DEA model proposed here is to measure the performance of the farmer's decision regarding the composition of the rearing production system. Thus, the DMUs are the 21 modal systems, identified from the panel discussions in 21 cities in seven Brazilian states.

7.2.2 Variables

The technicians and researchers mentioned in item 7.1, analysed the variables set and immediately identified those relevant to our study. They selected "number of bulls" as the input variable, since this variable represents a significant portion of all total expenditures of the ranchers that produce calves, being directly linked to the quality of animals that will be sold in these systems. This is also the only category that is purchased from other herd, especially in ranches with herds of genetic selection.

The products of the system that generate the main revenue from the cow-calf systems were chosen as the outputs variables. These are the "number of calves on the herd" and the "number of cull cows". All calves produced are sold on the property and generate income. Cull cows are those that are sold, as they are no more be part of the herd production system, either by higher age or by reproductive performance lower than desired.

The variables indicated by experts need to be examined by analysts to determine whether they conform to the properties required by the DEA models. In particular, there must be a causal relationship between each input-output pair (Gomes et al., 2009). There is a clear causal relationship between the output "number of calves on the herd" and the input "number of bulls". The same cannot be said of the relationship between the input and the output "number of cull cows". Actually, there is no direct causal relationship between these variables; however there is a cost-benefit relationship. In the case the rancher has a great number of bulls (that represent an expense) he must earn more, either through the sale of calves or cows. Therefore, the "bulls – cull cows" ratio makes sense when using DEA to analyze cost-benefit ratios, and not just pure productive relations. This interpretation of DEA was introduced by Womer et al. (2006) and was used by Kuosmanen & Kortelainen (2007), Kuosmanen et al. (2009). Generalizations of this usage can be seen in Bougnol et al. (2010). Table 9 presents the data. It appears that the herds are of different scales of production.

7.2.3 Model

In this chapter we use the DEA BCC model, since there was no evidence of proportionality between inputs and outputs, and the scales of production are known to differ between the modal systems.

We chose the input oriented model, since the objective is to evaluate the performance regarding the farmer's decision, which is based upon the purchase of bulls, on the most efficient use of the breeding animals kept on the herd, as a strategy to reduce costs.

7.3 Results

Table 10 shows the efficiency measurements and the multipliers based on the classic DEA BCC model. The value of u_* for DMU3, DMU5 and DMU8 are negative. When the multipliers of the outputs and the u_* are used to evaluate other DMUs, negative efficiencies will appear. Table 11 and 12 show the results based on the Modified DEA BCC model.

From Table 11 one can see that DMU3 and DMU5 that are efficient in the classic model are not in the modified DEA model. DMU3, Aquidauana was efficient by default, because it has the highest values for the outputs. Efficient by default DMUs may not be really efficient. Its

DMUs			Breeders (input)	Calves (output)	Cull cows (output)
Municipality	State	Code			
Alvorada	TO	DMU1	12	147	30
Amanbaí	MS	DMU2	15	143	40
Aquidauana	MS	DMU3	92	713	214
Bonito	MS	DMU4	14	166	75
Brasilândia	MS	DMU5	31	290	178
Camapuã	MS	DMU6	9	65	33
Carlos Chagas	MG	DMU7	19	297	160
Catalão	GO	DMU8	8	81	42
Corumbá	MS	DMU9	69	455	200
Itamarajú	BA	DMU10	4	44	18
Lavras do Sul	RS	DMU11	5	58	30
Montes Claros	MG	DMU12	5	47	28
Niquelândia	GO	DMU13	4	35	18
Paraíso do Tocantins	TO	DMU14	12	123	35
Porangatu	GO	DMU15	5	46	23
Ribas Rio Pardo	MS	DMU16	15	143	70
Rio Verde	GO	DMU17	23	196	82
São Gabriel d'Oeste	MS	DMU18	11	95	40
Tupã	SP	DMU19	5	46	30
Uberaba	MG	DMU20	5	66	36
Uberlândia	MG	DMU21	2	20	10

Table 9. DMUs, inputs and outputs.

efficiency may be all due to mathematical distortions. The use of the modified DEA BCC model may help to identify if an efficient by default DMU is really efficient. In our case study, Aquidauana loses its efficiency when using the proposed modified DEA BCC model. We can conclude that its efficiency in the DEA BCC model is due only to a mathematical distortion. DMU5 and DMU9 present a loss of efficiency, and although not efficient by default they are very close to this situation. Further studies are needed to better explain the figures for theses DMUs.

DMU	Multipliers				Eff	DMU	Multipliers				Eff
	v	u_1	u_2	u_*			v	u_1	u_2	u_*	
1	0.083	0.005	0.000	0.064	0.816	11	0.200	0.012	0.000	0.155	0.866
2	0.067	0.004	0.000	0.052	0.637	12	0.200	0.000	0.023	0.173	0.808
3	0.011	0.002	0.000	-0.360	1.000	13	0.250	0.015	0.000	0.193	0.730
4	0.071	0.004	0.000	0.055	0.783	14	0.083	0.005	0.000	0.064	0.693
5	0.032	0.000	0.022	-2.828	1.000	15	0.200	0.012	0.000	0.155	0.719
6	0.111	0.007	0.000	0.086	0.529	16	0.067	0.004	0.000	0.052	0.637
7	0.053	0.003	0.000	0.000	1.000	17	0.043	0.003	0.000	0.034	0.557
8	0.125	0.008	0.000	0.097	0.718	18	0.091	0.006	0.000	0.070	0.600
9	0.014	0.000	0.025	-3.922	0.990	19	0.200	0.000	0.023	0.173	0.853
10	0.250	0.015	0.000	0.193	0.868	20	0.200	0.000	0.023	0.173	0.989
						21	0.500	0.031	0.000	0.386	1.000

Table 10. Results based on the classic DEA BCC model, input oriented (Eff = Efficiency).

DMU	Multipliers				Eff	DMU	Multipliers				Eff
	v	u_1	u_2	u_*			v	u_1	u_2	u_*	
1	0.083	0.005	0.000	0.064	0.816	11	0.200	0.012	0.000	0.155	0.866
2	0.067	0.004	0.000	0.052	0.637	12	0.200	0.000	0.023	0.173	0.808
3	0.011	0.001	0.000	-0.015	0.517	13	0.250	0.015	0.000	0.193	0.730
4	0.071	0.004	0.000	0.055	0.783	14	0.083	0.005	0.000	0.064	0.693
5	0.032	0.000	0.004	-0.041	0.686	15	0.200	0.012	0.000	0.155	0.719
6	0.111	0.007	0.000	0.086	0.529	16	0.067	0.004	0.000	0.515	0.637
7	0.053	0.003	0.000	0.041	1.000	17	0.043	0.003	0.000	0.336	0.557
8	0.125	0.008	0.000	0.097	0.718	18	0.091	0.006	0.000	0.070	0.600
9	0.014	0.001	0.000	-0.020	0.432	19	0.200	0.000	0.023	0.173	0.853
10	0.250	0.015	0.000	0.193	0.868	20	0.200	0.000	0.023	0.173	0.989
						21	0.500	0.000	0.057	0.433	1.000

Table 11. Results based on the Modified DEA BCC model, input oriented (Eff = Efficiency).

Other DMUs present similar results in both models. DMU7 and DMU 21 (Carlos Chagas and Uberlândia) are efficient. Uberlândia has the lowest input value and so it is also an efficient by default DMU. As this DMU is located in the decreasing returns to scale region, the modified DEA BCC model does not help to decide if it is really efficient or not.

The production systems developed Carlos Chagas is a medium scale one. The reproductive indexes of the cow matrix are very good, reflecting the good husbandry with breeders' efficient use. That is, the system showed proportionally greater production of calves for sale, with a smaller number of bulls purchased (within the range of each system).

	1	2	3	4	5	6	7	8	9	10	11	12	13	14	15	16	17	18	19	20	21
1	0.816	0.637	0.484	0.783	0.599	0.529	1.000	0.718	0.416	0.868	0.866	0.731	0.730	0.693	0.719	0.637	0.557	0.600	0.719	0.965	1.000
2	0.816	0.637	0.484	0.783	0.599	0.529	1.000	0.718	0.416	0.868	0.866	0.731	0.730	0.693	0.719	0.637	0.557	0.600	0.719	0.965	1.000
3	0.726	0.563	0.517	0.716	0.598	0.343	1.001	0.523	0.433	0.412	0.522	0.371	0.258	0.589	0.357	0.563	0.525	0.468	0.357	0.632	0.000
4	0.816	0.637	0.484	0.783	0.599	0.529	1.000	0.718	0.416	0.868	0.866	0.731	0.730	0.693	0.719	0.637	0.557	0.600	0.719	0.965	1.000
5	0.211	0.253	0.281	0.588	0.686	0.324	1.000	0.507	0.349	0.253	0.507	0.456	0.253	0.264	0.329	0.507	0.397	0.345	0.507	0.659	0.000
6	0.816	0.637	0.484	0.783	0.599	0.529	1.000	0.718	0.416	0.868	0.866	0.731	0.730	0.693	0.719	0.637	0.557	0.600	0.719	0.965	1.000
7	0.816	0.637	0.484	0.783	0.599	0.529	1.000	0.718	0.416	0.868	0.866	0.731	0.730	0.693	0.719	0.637	0.557	0.600	0.719	0.965	1.000
8	0.816	0.637	0.484	0.783	0.599	0.529	1.000	0.718	0.416	0.868	0.866	0.731	0.730	0.693	0.719	0.637	0.557	0.600	0.719	0.965	1.000
9	0.726	0.562	0.517	0.715	0.597	0.343	1.000	0.523	0.432	0.411	0.521	0.370	0.257	0.589	0.357	0.562	0.525	0.468	0.357	0.631	0.000
10	0.816	0.637	0.484	0.783	0.599	0.529	1.000	0.718	0.416	0.868	0.866	0.731	0.730	0.693	0.719	0.637	0.557	0.600	0.719	0.965	1.000
11	0.816	0.637	0.484	0.783	0.599	0.529	1.000	0.718	0.416	0.868	0.866	0.731	0.730	0.693	0.719	0.637	0.557	0.600	0.719	0.965	1.000
12	0.356	0.360	0.273	0.669	0.679	0.512	1.000	0.703	0.341	0.727	0.853	0.808	0.727	0.403	0.695	0.587	0.442	0.491	0.853	0.989	1.000
13	0.816	0.637	0.484	0.783	0.599	0.529	1.000	0.718	0.416	0.868	0.866	0.731	0.730	0.693	0.719	0.637	0.557	0.600	0.719	0.965	1.000
14	0.816	0.637	0.484	0.783	0.599	0.529	1.000	0.718	0.416	0.868	0.866	0.731	0.730	0.693	0.719	0.637	0.557	0.600	0.719	0.965	1.000
15	0.816	0.637	0.484	0.783	0.599	0.529	1.000	0.718	0.416	0.868	0.866	0.731	0.730	0.693	0.719	0.637	0.557	0.600	0.719	0.965	1.000
16	0.816	0.637	0.484	0.783	0.599	0.529	1.000	0.718	0.416	0.868	0.866	0.731	0.730	0.693	0.719	0.637	0.557	0.600	0.719	0.965	1.000
17	0.816	0.637	0.484	0.783	0.599	0.529	1.000	0.718	0.416	0.868	0.866	0.731	0.730	0.693	0.719	0.637	0.557	0.600	0.719	0.965	1.000
18	0.816	0.637	0.484	0.783	0.599	0.529	1.000	0.718	0.416	0.868	0.866	0.731	0.730	0.693	0.719	0.637	0.557	0.600	0.719	0.965	1.000
19	0.356	0.360	0.273	0.669	0.679	0.512	1.000	0.703	0.341	0.727	0.853	0.808	0.727	0.403	0.695	0.587	0.442	0.491	0.853	0.989	1.000
20	0.356	0.360	0.273	0.669	0.679	0.512	1.000	0.703	0.341	0.727	0.853	0.808	0.727	0.403	0.695	0.587	0.442	0.491	0.853	0.989	1.000
21	0.356	0.360	0.273	0.669	0.679	0.512	1.000	0.703	0.341	0.727	0.853	0.808	0.727	0.403	0.695	0.587	0.442	0.491	0.853	0.989	1.000

Table 12. Cross evaluation matrix for the Modified DEA BCC model from the case study.

It is interesting to point out that in the study performed by Gomes et al. (2011), based on a DEA BCC model with weights restrictions and with the same dataset, Carlos Chagas and Uberlândia were the production systems that were more referenced as benchmarks (76% of non-zero contributions in the formation of the targets of the inefficient DMUs). The authors state that these modal systems can serve as a reference for the others, when assessing the performance of the cattle farmer's decision in relation to the production criteria. However, it is important to stress again that the DMU Uberlândia is efficient by default, and there must be some caution when indicating it as a benchmark.

One other advantage of the modified DEA BCC model is to allow the use of Cross Evaluation for the variable returns to scale situation as shown in Table 12. This matrix leads to the Cross Evaluation ranking shown in Table 13.

In Table 13 we can see that DMU3, Aquidauana, which was efficient with the original BCC model, has now one of the lowest efficiency measures when using the modified DEA BCC model and the cross evaluation technique. We can conclude that this DMU is a maverick, i.e. it is a false positive.

DMU	Municipality	Average Cross Efficiency
DMU7	Carlos Chagas	1.0000
DMU20	Uberaba	0.9230
DMU21	Uberlândia	0.8571
DMU11	Lavras do Sul	0.8139
DMU10	Itamarajú	0.7685
DMU4	Bonito	0.7455
DMU19	Tupã	0.7001
DMU12	Montes Claros	0.6985
DMU1	Alvorada	0.6910
DMU8	Catalão	0.6865
DMU13	Niquelândia	0.6617
DMU15	Porangatu	0.6614
DMU5	Brasilândia	0.6182
DMU16	Ribas Rio Pardo	0.6138
DMU14	Paraíso do Tocantins	0.6076
DMU2	Amanbaí	0.5586
DMU18	São Gabriel d'Oeste	0.5547
DMU17	Rio Verde	0.5241
DMU6	Camapuã	0.4983
DMU3	Aquidauana	0.4373
DMU9	Corumbá	0.4000

Table 13. Final ranking using the average cross evaluation index with the modified BCC model.

8. Concluding remarks

In this paper it was shown that the input orientated DEA BCC model can generate negative efficiencies that are usually hidden in the model. With the help of an example of one input and one output, the condition for the possible occurrence of negative efficiencies was shown. Furthermore it was shown that a small intuitive change in the BCC multipliers model avoids that situation.

The inclusion of a new set of restrictions in the BCC multipliers model generates an important modification in the dual of the envelope model. The new dual model, with a new set of variables, may change the efficient frontier.

Two numerical examples were presented in this article. The first example has two variables, one input and one output. It was geometrically shown that the modified model causes a change in the efficient frontier. These changes were interpreted by the introduction of a non-observed DMU. In this first example, the DMU causing negative efficiency in other DMUs has a reduction of its efficiency in the modified model. This reduction did not occur in the second example (two inputs and one output). This happened due to the optimum multipliers set multiplicity.

A real case study was carried out with the Modified DEA BCC model here proposed, regarding the evaluation of 21 Brazilian beef cattle modal production systems that performed only the cow-calf phase. The DMUs that could cause negative efficiencies were identified, and the new efficiencies were calculated. We could observe that most of the evaluated systems work with increasing returns to scale and lose efficiency. That is, they could produce more and adjusted the scale if they invest in balancing the number of bulls in the herds. The production systems in the municipalities of Carlos Chagas and Uberlândia are examples of this balance between investment in higher production of bulls and calves for sale, within ranges consistent with the income of ranchers.

It is very important to observe that the negative efficiencies mentioned in this paper apply neither to the CCR model nor to the output orientated BCC model. The latter possesses a restriction in the multipliers model that is of the "greater than" or "equal to" type, guaranteeing the non negativity of the efficiency measures.

Finally, another situation susceptible to the appearance of negative efficiencies in DEA context is when extending the MCDEA model (Li & Reeves, 1999) to the variable returns to scale assumption, especially if the MCDEA-TRIMAP efficiency is used (Soares de Mello et al., 2009). Future studies should verify the conditions in which the phenomenon of negative efficiencies occurs in the MCDEA model, and how to avoid them.

9. References

Angulo-Meza, L.; Biondi Neto, L.; Soares de Mello, J. C. C. B. & Gomes, E. G. (2005). ISYDS - Integrated System for Decision Support (SIAD Sistema Integrado de Apoio a Decisão): A Software Package for Data Envelopment Analysis Model. *Pesquisa Operacional*, Vol. 25, No. 3, pp. 493-503

Angulo-Meza, L.; Soares de Mello, J. C. C. B.; Gomes, E. G. & Biondi Neto, L. (2004). Eficiências negativas em modelos DEA-BCC: como surgem e como evitá-las. *VII Simpósio de Pesquisa Operacional e Logística da Marinha - SPOLM*, Niterói.

Banker, R. D.; Charnes, A. & Cooper, W. W. (1984). Some models for estimating technical scale inefficiencies in data envelopment analysis. *Management Science*, Vol. 30, No. 9, pp. 1078-1092, 0025-1909

Bougnol, M. L.; Dulá, J. H.; Estellita Lins, M. P. & Moreira da Silva, A. C. (2010). Enhancing standard performance practices with DEA. *Omega*, Vol. 38, No. 1-2, pp. 33-45

Centro de Estudos Avançados em Economia Aplicada (2010). *Metodologia do índice de preços dos insumos utilizados na produção pecuária brasileira.* Available at: <http://www.cepea.esalq.usp.br/boi/metodologiacna.pdf >. Accessed: 24 March 2010.

Charnes, A.; Cooper, W. W. & Rhodes, E. (1978). Measuring the efficiency of decision-making units. *European Journal of Operational Research*, Vol. 2, pp. 429-444, 0377-2217.

Cooper, W. W.; Seiford, L. M. & Zhu, J. (2004). *Handbook on Data Envelopment Analysis*, Kluwer Academic Publishers, Boston.

Doyle, J. & Green, R. H. (1994). Efficiency and cross-efficiency in DEA derivations, meanings and uses. *Journal of the Operational Research Society*, Vol. 45, pp. 567-578

Duzakin, E. & Duzakin, H. (2007). Measuring the performance of manufacturing firms with super slacks based model of data envelopment analysis: An application of 500 major industrial enterprises in Turkey. *European Journal of Operational Research*, Vol. 182, No. 3, pp. 1412-1432

Farrell, M. J. (1957). The Measurement of Productive Efficiency. *Journal of Royal Statistical Society Series A*, Vol. 120, No. 3, pp. 253-281

Gomes, E. G.; Abreu, U. G. P. d.; Soares de Mello, J. C. C. B.; Carvalho, T. B. & Zen, S. (2011). DEA performance evaluation of livestock systems in Brazil *International Conference on Data Envelopment Analysis and Its Applications to Management.*, Lima.

Gomes, E. G.; Soares de Mello, J. C. C. B.; Souza, G. d. S.; Angulo-Meza, L. & Mangabeira, J. A. d. C. (2009). Efficiency and sustainability assessment for a group of farmers in the Brazilian Amazon. *Annals of Operations Research*, Vol., pp.

Iqbal Ali, A. & Seiford, L. M. (1990). Translation invariance in data envelopment analysis. *Operations Research Letters*, Vol. 9, No. 6, pp. 403-405

Kuosmanen, T.; Bijsterbosch, N. & Dellink, R. (2009). Environmental cost-benefit analysis of alternative timing strategies in greenhouse gas abatement: A data envelopment analysis approach. *Ecological Economics*, Vol. 68, No. 6, pp. 1633-1642

Kuosmanen, T. & Kortelainen, M. (2007). Valuing environmental factors in cost-benefit analysis using data envelopment analysis. *Ecological Economics*, Vol. 62, No. 1, pp. 56-65

Li, X.-B. & Reeves, G. R. (1999). A multiple criteria approach to data envelopment analysis. *European Journal of Operational Research*, Vol. 115, No. 3, pp. 507-517

Lins, M. P. E. & Angulo-Meza, L. (2000). *Análise Envoltória de Dados e perspectivas de integração no ambiente de Apoio à Decisão*, Editora da COPPE/UFRJ, Rio de Janeiro.

Lins, M. P. E.; Angulo-Meza, L. & Moreira da Silva, A. C. (2004). A multi-objective approach to determine alternative targets in data envelopment analysis. *Journal of the Operational Research Society*, Vol. 55, No. 10, pp. 1090-1101

Nacif, F. B.; Soares de Mello, J. C. C. B. & Angulo-Meza, L. (2009). Choosing weights in optimal solutions for DEA-BCC models by means of a n-dimensional smooth frontier. *Pesquisa Operacional*, Vol. 29, No. 3, pp. 623-642

Pastor, J. T. (1996). Translation invariance in data envelopment analysis: A generalization. *Annals of Operations Research*, Vol. 66, pp. 93-102

Plaxico, J. S. & Tweeten, L. G. (1963). Representative farms for policy and projection research. *Journal of Farm Economics,* Vol. 45, pp. 1458-1465

Portela, M. C. A. S.; Thanassoulis, E. & Simpson, G. (2004). Negative data in DEA: A directional distance approach applied to bank branches. *Journal of the Operational Research Society,* Vol. 55, No. 10, pp. 1111-1121

Quariguasi Frota Neto, J. & Angulo-Meza, L. (2007). Alternative targets for data envelopment analysis through multi-objective linear programming: Rio de Janeiro Odontological Public Health System Case Study. *Journal of the Operational Research Society,* Vol. 58, pp. 865–873

Rosen; Schaffnit, C. & Paradi, J. C. (1998). Marginal rates and two dimensional level curves in DEA. *Journal of Productivity Analysis,* Vol. 9, No. 3, pp. 205-232

Sexton, T. R.; Silkman, R. H. & Logan, A. J. (1986). Data Envelopment Analysis: Critique and extensions. IN Silkman, H. (Ed.) *Measuring efficiency: An assessment of data envelopment analysis.* Jossey-Bass Editor, San Francisco.

Sharp, J. A.; Meng, W. & Liu, W. (2007). A modified slacks-based measure model for data envelopment analysis with 'natural' negative outputs and inputs. *Journal of the Operational Research Society,* Vol. 58, No. 12, pp. 1672-1677

Soares de Mello, J. C. C. B.; Climaco, J. C. N. & Angulo-Meza, L. (2009). Efficiency evaluation of a small number of DMUs: an approach based on Li and Reeves's model. . *Pesquisa Operacional,* Vol. 29, Issue 1, pp. 97-110.

Soares de Mello, J. C. C. B.; Lins, M. P. E. & Gomes, E. G. (2002). Construction of a smoothed DEA frontier. *Pesquisa Operacional,* Vol. 28, No. 2, pp. 183-201

Souza, G. d. S. (2006). Significância de Efeitos Técnicos na Eficiencia de Produção da Pesquisa Agropecuária Brasileira. *Revista Brasileira de Economia - RBE,* Vol. 60, No. 1, pp. 94-117

Sueyoshi, T. (2004). Mixed integer programming approach of extended DEA-discriminant analysis. *European Journal of Operational Research,* Vol. 152, No. 1, pp. 45-55

Thanassoulis, E. & Allen, R. (1988). Simulating Weight restrictions in Data Envelopment Analysis by means of unobserved DMUs. *Management Science,* Vol. 44, pp. 586-594.

Thrall, R. M. (1996). The lack of invariance of optimal dual solutions under translation. *Annals of Operations Research,* Vol. 66, pp. 103-108

Womer, N. K.; Bougnol, M. L.; Dulá, J. H. & Retzlaff-Roberts, D. (2006). Benefit-cost analysis using data envelopment analysis. *Annals of Operations Research,* Vol. 145, No. 1, pp. 229-250

Wu, J.; Liang, L. & Chen, Y. (2009). DEA game cross-efficiency approach to Olympic rankings. *Omega,* Vol. 37, No. 4, pp. 909-918

8

A Ranking for the Vancouver 2010 Winter Olympic Games Based on a Hierarchical Copeland Method

João Carlos Correia Baptista Soares de Mello
and Nissia Carvalho Rosa Bergiante
Universidade Federal Fluminense
Brazil

1. Introduction

Many studies have been carried out to evaluate the results of the Olympic Games. Some of them are based on how these events can bring benefits to the host cities [Glynn (2008); Cheng (2008); Xiaoduo and Jianxin (2008)] and others are interested in social studies [Bernstein (2000), Farrell (1989), Levine (1974) and Ball (1972)].

Besides these studies, we can find researches in the environmental and health areas [Hadjichristodoulou, Mouchtouri and Vaitsi (2006); Allen et al (2006); Weiler, Layton and Hunt (1998); Streets et al (2007)], some about tourism industry and others that evaluate mathematics and economics aspects of the Games [Heazlewood (2006), Bernard and Busse (2004), Lins et al (2003) and Li et al (2008)].

In this work we are interested in studying the results of the Winter Olympic Games, held in Vancouver in 2010, in order to propose a new ranking to the countries that took part in the Games. Traditionally, as discussed in Soares de Mello et al (2009), the International Olympic Committee (IOC) shows the results of both Summer and Winter Olympic Games in a table that rank the nations by the number of gold medals won. This is a Lexicographic method. Although this is not an official ranking, to the media and people in general, this is a summary of the participation of the nations, i.e., their production in the games.

In the literature we can find other studies that suggest alternative rankings for the Olympic Games. As an alternative to rankings based on Lexicographic Methods, some studies as Lozano et al (2002) use Data Envelopment Analysis models with population and GNP as inputs and the number of gold, silver and bronze medals as outputs, to rank the nations. Others authors as Lins et al (2003) improved that model by adding a new constraint that defined as a constant the total number of medals (Zero Sum Gains DEA model (ZSG-DEA).

As an extension of these studies some authors use others social and economics variables to build their models such as Churilov and Flitman (2006). These results found in Lozano et al (2002) were followed by the proposal of using GDP per capita and not only GDP as an input and creating weight restriction to the DMUs as done by Li et al (2008). In almost a sequence many researchers have been published new approaches to the same problem [Wu et al

(2008); Wu, Liang and Chen (2009); Wu, Liang and Yang (2009); Yang et al (2009) and Zhang et al (2009)].

Besides all the works showed before, we can mention studies that took into account only the number of medals won by each country. Soares de Mello et al (2008) used as input the number of athletes brought to the country to the game and Bergiante and Soares de Mello (2010) proposed as output, the number of medals won by each country. There are other works that evaluated the home advantage event [Balmer, Nevill and Williams (2001) and Balmer, Nevill and Williams (2003)] and the difference between the Summer and Winter Olympic Games [Johnson and Ali (2004)]. Beyond that, some rankings are built taking into account the total number of medals, the Lexicographic method (explained by Lins et al (2003)), which under evaluate the gold medal, and others range countries only by the number of gold medals, which don't evaluate the results of silver and bronze medals.

Although the importance of all these studies to the advance of the knowledge of the Olympic Games we noted that a few of them, as can be seen in Soares de Mello et al (2009), concerned about the difference in the value of a medal when comparing different modalities, for example, ice hockey has, at least, five times to get a medal, while, to win a biathlon sport, an athlete has to compete only one time.

Due to this important aspect, in this paper, we propose the use of ordinal multicriteria methods to study the results of the Vancouver 2010 Olympic Games. In our case we intended to use ordinals methods as Condorcet and Copeland [Barba-Romero and Pomerol (1997)] to hierarchy the nations.

To do so, we will divide our study in two steps. In the first part, by gathering information about how many medals each country won in each modality, we will establish two kinds of ranking. One is taking into account the total number of medals per modalities, and the second one is regarding only the number of gold medals distributed by modalities. The idea to use these two rankings, not only one, is based on the criticism of some studies as Soares de Mello et al (2008) that argues that the first type of ranking, based on the total number of medals, underestimate the gold medals and the second ranking, based on the number of gold medals, over evaluate the gold medals.

After built these two rankings, we are going to apply the Copeland method, using them as criteria. As a result, we aim to establish a ranking per each modality.

In the second step, each ranking built in the earlier stage will be seen as criteria to a new application of the Copeland Method to establish our final ranking. As one can note, this final ranking will take into account three different models to assort the countries: the Lexicographic method, the total number of medals and the difference between all the modalities of the Olympic Games. As far as we know, this proposal is an advance in the studies in this field. This can be seen as a hierarchy Copeland method.

It is important to point out the utility of the studies this one in terms of operations management. As discussed by Roy (1992), multicriteria methods can optimize the decision process through the study of the value of each alternative or action taken by the decision maker and the analysis of their robustness. In fact, all these approaches contribute to the efficiency of the operations management processes in many different scenarios, as companies, events as Olympic Games and so on.

By improving the operational efficiency these methods help organizations to accomplish their objectives, strategies and action plans. They also allow a better control of the resources involved in the business. For all kind of organizations, managing operations remains a priority and because of that, multicriteria method should be seen as an useful tool to operations managers and operations executives since they enhance the process of decision making.

The next section will point out some aspects of the Winter Olympic Game. The section 3 summarizes the Methodology used, i.e., the multicriteria decision support tools and in the Section subsequent we described the model used in this study. After that, we will present the results and a briefly analysis of them. Following we summarize our conclusions and some future research directions.

2. The winter olympic games

In 776 AC, in Olympia, Greece, was born the Ancient Games. As an evolution of this proposal, in 1896, it started the Modern Games, as discussed by Wallechinsky (2004). But, until 1920, the only competition existing was the Summer Olympic Games. The Winter version was held in Chamonix, France in 1924, and since then, these two competitions have been happened every four year, in the same year. However, from 1994 and on, these games have been staggered two years apart.

Some authors, as Johnson and Ali (2004), have been study the difference between these two Olympic Games. These researchers found that the ability to participate in both Games is not the same to all countries and even if all the countries took part in the two Games, they will not have an equal ability to win medals [Bergiante and Soares de Mello (2010)]. This study argues, as expected, that countries with heavy winter will have better results in the Winter Games than in the Summer Games. Another research of Balmer, Nevill and Williams (2001) found home advantages in some modalities in the Winter Games related to familiarity with local conditions which could prejudice away athletes.

In the very first Winter Olympic Games, in 1924, there were a total of 258 athletes representing 16 countries and six modalities to compete. The majority of the nation that took part in the Games was from Europe and North America.

Nowadays, the participation in these Games has been enlarged. The last competition, held in Vancouver, Canada in 2010, is an example of this. A total of 82 countries, including nations from South America, as Brazil, and others as India and Hong Kong, attended to the Games. There were a total of 2600 athletes participated in the events, which is much more than the 258 of Chamonix, France, 1924.

In relation to the number of disciplines programmed, there was a huge improvement. In the past, as mentioned before, there were only six sports: bobsleigh, curling, ice hockey, figure and speed skating, skiing and the military patrol race. In the 2010 version of the Winter Olympic Games, a total of fifteen sports were programmed. They were: alpine skiing, biathlon, bobsleigh, cross-country, curling, figure skating, freestyle skiing, ice hockey, luge, nordic combined, short track, skeleton, ski jumping, snowboard and speed skating.

Obviously, the increment also happens in the number of medals distributed in these Games. Comparing to the first Game, in which 49 medal were disputed, Vancouver, 2010 Games, had more than five times medals to distributed, adding up a total of 258 medals (including gold, silver and bronze). Since many competitions were composed by teams, an amount of 615 medals was awarded.

In terms of medals, Johnson and Ali (2004), in the same work presented before, will argue that it has become easier to win any medal as the number of available medals in the Olympic Games has increased. From this only study we could infer that win a medal in the Winter Game is harder than in the Summer Game. However, Bergiante and Soares de Mello (2010) discussed, based on the work of Hilvoorde, Elling and Stokvis (2010), that due to the obtainment of the medals and all their positive impact as economy growth and a superior international prestige, many countries have been invested more and more to achieve a better position in the medals ranking in the Olympic Games.

So, although the number of medal has increased, the number of countries that took part in the games has also growth. This become clear the difficulty to win a medal in the Olympic Games and, as might as be expected, it was not different in the Winter Olympic Games.

3. Borda, Condorcet and Copeland methods

In situations where there are a lot of criteria and objectives, some conflicting to analyze a problem, it is convenient to use a Multicriteria Decision Support. This system consists in methods and techniques to help the decision making process [Roy and Bouyssou (1993)].

There are many multicriteria decision methods. In this work we are interested in the ordinals methods. Some authors as Gomes et al (2009) affirm that there are three most referenced ordinal methods namely Borda, Condorcet and Copeland methods. The authors said that these methods are seen as intuitive and even less complex in terms of computational efforts and information needed to solve the problems.

In these methods the data requested to the decision makers is, based on their preferences, the ranking of alternatives with a preorder for each criterion [Barba-Romero and Pomerol (1997)]. The Borda method was proposed by Jean-Charles de Borda in the 18th century and it is used to aggregate binary relations among the alternatives. An ordinal scale is given to the decision makers. They must order the alternatives based on their preferences, attributing a certain number of points for each first, second and other places of the ranking.

If there are n alternatives to be ranked, the alternative preferred is worth 1 point, the second place vote is worth 2 points, and so on all the way to an nth place vote, which will worth the lesser pointing.

In the end, these points are counted and the alternatives are sorting by the crescent order of pointing (respecting the totality axiom). The first alternative considered will be that one with the lesser sum of points.

Despite the simplicity of the Borda Method, some authors as Gomes et al (2009) argued that this method fails to satisfy the Arrow axiom [Arrow (1951)], i.e., the independence of irrelevant alternatives. Wherefore, the final ranking would be relative to the group of alternatives evaluated which is not a desirable situation.

In the Condorcet Method, the decision makers are required to express their preferences in a series of pairwise comparisons. From that information, it is possible to build a graph to express their relations [Boaventura Neto (2003)].

Using the graph built we establish preferences relations of the alternatives. We will select a dominant alternative, i.e., an alternative that beats every other feasible alternative in the pairwise comparison. The decision procedure of comparing pairs of alternatives might lead to an intransitive situation, in which Condorcet winners may not exist.

In the Condorcet paradox, as this situation is called, any alternative can be reached from any other by a sequence of alternatives. An example, called Condorcet triplet, can be illustrated as follows: an alternative A beats the alternative B, and B beats C. However, the alternative C beats A leading to a cycle where a Condorcet winner cannot exist. When the intransitive situation does not happen, the Condorcet Method should be used instead of Borda Method [Soares de Mello, Quintella and Soares de Mello (2004)].

A preference aggregation method called Copeland method was design to overcome the voting cycles that impedes to determine a Condorcet winner. The Copeland method uses the adjacency matrix of the Condorcet method graph. After the pairwise comparison, we calculate the number of simply majority wins minus losses, for each alternative. Hence, the Copeland ranking is given by ordering the alternatives according to the results of this sum.

The Copeland method always gives an order of alternatives and in its set will contain the Condorcet winner if there is not an intransitive cycle. Its computational effort is higher than the Condorcet method but it is able to reduce the influence of the irrelevant alternatives [Gomes Junior, Soares de Mello and Soares de Mello (2008)].

Due to all the arguments presented before, in this work we choose to use the Copeland method since its taking into account important issues of the Borda and Condorcet approaches.

The method proposed here has two important goals. The first one is to propose a ranking that will not overvalue the gold medal as generally happens in the Lexicographic method. The second one is to use an approach that take into account that are differences in terms of value among all the disciplines and their medals in the Olympic Games. As an example in the Alpine Skiing there are five possibilities to earn a gold medal but in Ice Hockey there are only two opportunities (men's and women's tournaments).

There are others studies that try to take into consideration these aspects and some of them use DEA methods, as Soares de Mello et al (2009) and Lins et al (2003). However we believe that our approach truly valued the differences among all the medals disputed and fundamentally with less effort than in the DEA methods.

In the second step, each ranking built in the earlier stage will be seen as criteria to a new application of the Copeland Method to establish our final ranking. As one can note, this final ranking will take into account three different models to assort the countries: the Lexicographic method, the total number of medals and the difference between all the modalities of the Olympic Games. As far as we know, this proposal is an advance in the studies in this field. This can be seen as a hierarchy Copeland method.

4. Modelling

As we want to propose a new ranking to the 2010 Winter Olympic Games, in this section we are going to fix up the elements of the problem. Some authors as Gomes et al (2009) argue that to organize a multicriteria problem, we should define the alternatives, the criteria and choose an appropriate method to solve the problem.

Based on the explanation in the third section, we are going to use the Copeland Method, applied in two phases. To the first step, the alternatives will be all the countries that took part in the Vancouver Winter Games, gathered by the modalities disputed. A complete example of data used in the first stage is given in Table 1. The next five tables are excerpts of others modalities.

ALPINE SKIING				
COUNTRY	GOLD MEDALS	SILVER MEDALS	BRONZE MEDALS	RANKING TOTAL NUMBER OF MEDALS	RANKING NUMBER OF GOLD MEDALS
United States	2	3	3	1	2
Germany	1	2	1	2	4
Canada	1	1	2	2	5
Norway	3	0	0	4	1
Austria	2	0	1	4	3
Russia	0	2	0	6	7
South Korea	0	2	0	6	7
China	0	0	2	6	9
Sweden	1	0	0	9	6
France	0	0	1	9	10
Switzerland	0	0	0	11	11
Netherlands	0	0	0	11	11
Czech Republic	0	0	0	11	11
Poland	0	0	0	11	11
Italy	0	0	0	11	11
Japan	0	0	0	11	11
Finland	0	0	0	11	11
Australia	0	0	0	11	11
Belarus	0	0	0	11	11
Slovakia	0	0	0	11	11
Croatia	0	0	0	11	11
Slovenia	0	0	0	11	11
Latvia	0	0	0	11	11
Great Britain	0	0	0	11	11
Kazakhstan	0	0	0	11	11
Estonia	0	0	0	11	11
Albania	0	0	0	11	11
Algeria	0	0	0	11	11
Andorra	0	0	0	11	11
Argentina	0	0	0	11	11
Armenia	0	0	0	11	11
Azerbaijan	0	0	0	11	11
Belgium	0	0	0	11	11
Bermuda	0	0	0	11	11
Bosnia & Herzegovina	0	0	0	11	11
Brazil	0	0	0	11	11
Bulgaria	0	0	0	11	11

ALPINE SKIING				
COUNTRY	GOLD MEDALS	SILVER MEDALS	BRONZE MEDALS	RANKING TOTAL NUMBER OF MEDALS	RANKING NUMBER OF GOLD MEDALS
Cayman Islands	0	0	0	11	11
Chile	0	0	0	11	11
Chinese Taipei	0	0	0	11	11
Colombia	0	0	0	11	11
Costa Rica	0	0	0	11	11
Cyprus	0	0	0	11	11
Denmark	0	0	0	11	11
North Korea	0	0	0	11	11
Ethiopia	0	0	0	11	11
Macedonia	0	0	0	11	11
Georgia	0	0	0	11	11
Ghana	0	0	0	11	11
Greece	0	0	0	11	11
Hong Kong	0	0	0	11	11
Hungary	0	0	0	11	11
Iceland	0	0	0	11	11
India	0	0	0	11	11
Iran	0	0	0	11	11
Ireland	0	0	0	11	11
Israel	0	0	0	11	11
Jamaica	0	0	0	11	11
Kenya	0	0	0	11	11
Kyrgyzstan	0	0	0	11	11
Lebanon	0	0	0	11	11
Liechtenstein	0	0	0	11	11
Lithuania	0	0	0	11	11
Mexico	0	0	0	11	11
Moldova	0	0	0	11	11
Monaco	0	0	0	11	11
Mongolia	0	0	0	11	11
Montenegro	0	0	0	11	11
Morocco	0	0	0	11	11
Nepal	0	0	0	11	11
New Zealand	0	0	0	11	11
Pakistan	0	0	0	11	11
Peru	0	0	0	11	11
Portugal	0	0	0	11	11
Serbia	0	0	0	11	11
Romania	0	0	0	11	11
San Marino	0	0	0	11	11
Senegal	0	0	0	11	11
South Africa	0	0	0	11	11
Spain	0	0	0	11	11
Tajikistan	0	0	0	11	11
Turkey	0	0	0	11	11
Ukraine	0	0	0	11	11
Uzbekistan	0	0	0	11	11

Table 1. Example of the ranking per modality (including all countries) - Alpine Skiing.

BIATHLON

COUNTRY	GOLD MEDALS	SILVER MEDALS	BRONZE MEDALS	RANKING TOTAL NUMBER OF MEDALS	RANKING NUMBER OF GOLD MEDALS
France	1	2	3	6	1
Norway	3	2	0	5	2
Germany	2	1	2	5	2
Russia	2	1	1	4	4
Slovakia	1	1	1	3	5
Austria	0	2	0	2	6
Belarus	0	1	1	2	6
Sweden	1	0	0	1	8
Kazakhstan	0	1	0	1	8
Croatia	0	0	1	1	8

Table 2. Example of ranking per modality – Biathlon.

CROSS-COUNTRY SKIING

COUNTRY	GOLD MEDALS	SILVER MEDALS	BRONZE MEDALS	RANKING TOTAL NUMBER OF MEDALS	RANKING NUMBER OF GOLD MEDALS
Norway	5	2	2	9	1
Sweden	3	2	2	7	2
Germany	1	4	0	5	3
Russia	1	1	2	4	4
Poland	1	1	1	3	5
Czech Republic	0	0	2	2	6
Finland	0	0	2	2	6
Switzerland	1	0	0	1	8
Estonia	0	1	0	1	8
Italy	0	1	0	1	8

Table 3. Example of ranking per modality – Cross-Country Skiing.

FREE STYLE SKIING

COUNTRY	GOLD MEDALS	SILVER MEDALS	BRONZE MEDALS	RANKING TOTAL NUMBER OF MEDALS	RANKING NUMBER OF GOLD MEDALS
United States	1	1	2	4	1
Canada	2	1	0	3	2
China	0	1	2	3	2
Australia	1	1	0	2	4
Norway	0	1	1	2	4
Belarus	1	0	0	1	6
Switzerland	1	0	0	1	6
Austria	0	1	0	1	6
France	0	0	1	1	6

Table 4. Example of ranking per modality – Free Style Skiing.

SNOWBOARD

COUNTRY	GOLD MEDALS	SILVER MEDALS	BRONZE MEDALS	RANKING TOTAL NUMBER OF MEDALS	RANKING NUMBER OF GOLD MEDALS
United States	2	1	2	5	1
Canada	2	1	0	3	2
France	0	1	2	3	2
Austria	0	1	1	2	4
Australia	1	0	0	1	5
Netherlands	1	0	0	1	5
Finland	0	1	0	1	5
Russia	0	1	0	1	5
Switzerland	0	0	1	1	5

Table 5. Example of ranking per modality – Snowboard.

Using the data of the Table 1 and all the others tables, we built the adjacency matrix by the Condorcet method. To the examples given before, as the Alpine Skiing modality, the matrices are shown hereafter. The last columns is calculated as showed here: ($\sum$ wins - $\sum$ losses)

ALPINE SKIING	United States	Norway	Austria	Germany	Switzerland	Croatia	Slovenia	Sweden	Italy	Czech Republic	Sum of Wins	Wins minus Losses
United States		1	1	1	1	1	1	1	1	1	9	8
Norway	0		1	1	1	1	1	1	1	1	8	5
Austria	0	0		1	1	1	1	1	1	1	7	3
Germany	1	1	1		1	1	1	1	1	1	9	6
Switzerland	0	1	1	0		1	1	1	1	1	7	3
Croatia	0	0	0	0	0		1	1	1	1	4	-3
Slovenia	0	0	0	0	0	1		1	1	1	4	-4
Sweden	0	0	0	0	0	0	1		1	1	3	-5
Italy	0	0	0	0	0	1	1	1		1	4	-4
Czech Republic	0	0	0	0	0	0	0	0	0		0	-1
Sum of losses	1	3	4	3	4	7	8	8	8	9		

1

Table 6. Adjacency matrix built by the Condorcet Method – Alpine Skiing.

BIATHLON	France	Norway	Germany	Russia	Slovakia	Austria	Belarus	Sweden	Kazakhstan	Croatia	Sum of Wins	Wins minus Losses
France		1	1	1	1	1	1	1	1	1	9	6
Norway	0		1	1	1	1	1	1	1	1	9	8
Germany	0	0		1	1	1	1	1	1	1	8	6
Russia	0	0	0		1	1	1	1	1	1	7	4
Slovakia	0	0	0	0		1	1	1	1	1	5	1
Austria	0	0	0	0	0		1	1	1	1	4	-2
Belarus	0	0	0	0	0	0		1	1	1	3	-4
Sweden	0	0	0	0	0	1	1		1	1	4	-3
Kazakhstan	0	0	0	0	0	0	0	0		1	1	-7
Croatia	0	0	0	0	0	0	0	0	0		0	-9
Sum of losses	3	1	2	3	4	6	7	7	8	9		

Table 7. Adjacency matrix built by the Condorcet Method – Biathlon.

CROSS-COUNTRY SKIING	Norway	Sweden	Germany	Russia	Poland	Czech Republic	Finland	Switzerland	Estonia	Italy	Slovenia	Sum of Wins	Wins minus Losses
Norway		1	1	1	1	1	1	1	1	1	1	10	10
Sweden	0		1	1	1	1	1	1	1	1	1	9	8
Germany	0	0		1	1	1	1	1	1	1	1	8	6
Russia	0	0	0		1	1	1	1	1	1	1	7	4
Poland	0	0	0	0		1	1	1	1	1	1	6	2
Czech Republic	0	0	0	0	0		1	1	1	1	1	5	-4
Finland	0	0	0	0	0	1		1	1	1	1	5	-4
Switzerland	0	0	0	0	0	1	1		1	1	1	5	-2
Estonia	0	0	0	0	0	1	1	0		1	1	4	-5
Italy	0	0	0	0	0	1	1	0	1		1	4	-5
Slovenia	0	0	0	0	0	0	0	0	0	0		0	-10
Sum of losses	0	1	2	3	4	9	9	7	9	9	10		

Table 8. Adjacency matrix built by the Condorcet Method - Cross-Country Skiing.

FREE STYLE SKIING	United States	Canada	China	Australia	Norway	Belarus	Switzerland	Austria	France	Sum of Wins	Wins minus Losses
United States		1	1	1	1	1	1	1	1	8	7
Canada	1		1	1	1	1	1	1	1	8	7
China	0	0		1	1	1	1	1	1	6	1
Australia	0	0	1		1	1	1	1	1	6	3
Norway	0	0	0	0		1	1	1	1	4	-2
Belarus	0	0	1	0	1		1	1	1	5	-1
Switzerland	0	0	1	0	1	1		1	1	5	-1
Austria	0	0	0	0	0	0	0		1	1	-6
France	0	0	0	0	0	0	0	0		0	-8
Sum of losses	1	1	5	3	6	6	6	7	8		

Table 9. Adjacency matrix built by the Condorcet Method – Free Style Skiing.

SNOWBOARD	United States	Canada	France	Austria	Australia	Netherlands	Finland	Russia	Switzerland	Sum of Wins	Wins minus Losses
United States		1	1	1	1	1	1	1	1	8	8
Canada	0		1	1	1	1	1	1	1	7	6
France	0	0		1	1	1	1	1	1	6	2
Austria	0	0	0		1	1	1	1	1	5	0
Australia	0	0	1	1		1	1	1	1	6	1
Netherlands	0	0	1	1	1		1	1	1	6	1
Finland	0	0	0	0	0	0		1	1	2	-5
Russia	0	0	0	0	0	0	1		1	2	-5
Switzerland	0	0	0	0	0	0	0	0		0	-8
Sum of losses	0	1	4	5	5	5	7	7	8		

Table 10. Adjacency matrix built by the Condorcet Method – Snowboard.

Now we are able to rank the countries and the results are seen in the following tables.

ALPINE SKIING	TOTAL ($\sum$ wins - $\sum$ losses)	RANKING
United States	8	1
Germany	6	2
Norway	5	3
Austria	3	4
Switzerland	3	4
Czech Republic	-1	6
Croatia	-3	7
Slovenia	-4	8
Italy	-4	8
Sweden	-5	10

Table 11. Ranking in the first stage – Alpine Skiing.

BIATHLON	TOTAL ($\sum$ wins - $\sum$ losses)	RANKING
Norway	8	1
France	6	2
Germany	6	2
Russia	4	4
Slovakia	1	5
Austria	-2	6
Sweden	-3	7
Belarus	-4	8
Kazakhstan	-7	9
Croatia	-9	10

Table 12. Ranking in the first stage – Biathlon.

CROSS-COUNTRY SKIING	TOTAL ($\sum$ wins - $\sum$ losses)	RANKING
Norway	10	1
Sweden	8	2
Germany	6	3
Russia	4	4
Poland	2	5
Switzerland	-2	6
Czech Republic	-4	7
Finland	-4	7
Estonia	-5	9
Italy	-5	9
Slovenia	-10	11

Table 13. Ranking in the first stage – Cross-Country Skiing.

FREESTYLE SKIING	TOTAL ($\sum$ wins - $\sum$losses)	RANKING
United States	7	1
Canada	7	1
Australia	3	3
China	1	4
Belarus	-1	5
Switzerland	-1	5
Norway	-2	7
Austria	-6	8
France	-8	9
United States	7	1
Canada	7	1

Table 14. Ranking in the first stage – Freestyle Skiing.

SNOWBOARD	TOTAL ($\sum$ wins - $\sum$losses)	RANKING
United States	8	1
Canada	6	2
France	2	3
Australia	1	4
Netherlands	1	4
Austria	0	6
Finland	-5	7
Russia	-5	7
Switzerland	-8	9
United States	8	1
Canada	6	2

Tabela 15. Ranking in the first stage – Snowboard.

We have obtained a ranking as in Table 11, Table 12, Table 13, Table 14 and Tabela 15 for all the 15 modalities in the Vancouver 2010 Olympic Games. These rankings were used as criteria to the second step of the method used here. An example of the aggregated results is showed in Table 16.

Position in the ranking by modality

	ALPINE SKIING	BIATHLON	BOBSLEIGH	CROSS-COUNTRY SKIING	CURLING	FIGURE SKATING	FREESTYLE SKIING	ICE HOCKEY	LUGE	NORDIC COMBINED	SHORT TRACK	SKELETON	SKI JUMPING	SNOWBOARD	SPEEDSKATING
United States	1	11	3	12	6	1	1	2	5	1	3	6	5	1	6
Germany	2	2	1	3	6	7	10	4	1	4	6	1	3	10	4
Canada	11	11	1	12	1	3	1	1	5	6	3	1	5	2	3
Norway	3	1	5	1	3	8	7	4	5	6	6	6	4	10	9
Austria	4	6	5	12	6	8	8	4	2	2	6	6	1	6	12
Switzerland	4	11	5	6	4	8	5	4	5	6	6	6	1	9	12
Russia	11	4	4	4	6	4	10	4	5	6	6	5	5	7	12
China	11	11	5	12	4	1	4	4	5	6	2	6	5	10	9
South Korea	11	11	5	12	6	4	10	4	5	6	1	6	5	10	1
France	11	2	5	12	6	8	9	4	5	3	6	6	5	3	12
Sweden	10	7	5	2	2	8	10	4	5	6	6	6	5	10	12
Netherlands	11	11	5	12	6	8	10	4	5	6	6	6	5	4	1
Czech Republic	6	11	5	7	6	8	10	4	5	6	6	6	5	10	5
Australia	11	11	5	12	6	8	3	4	5	6	6	6	5	4	12
Poland	11	11	5	5	6	8	10	4	5	6	6	6	2	10	9
Italy	8	11	5	9	6	8	10	4	4	4	5	6	5	10	12
Japan	11	11	5	12	6	4	10	4	5	6	6	6	5	10	7
Finland	11	11	5	7	6	8	10	3	5	6	6	6	5	7	12
Belarus	11	8	5	12	6	8	5	4	5	6	6	6	5	10	12
Slovakia	11	5	5	12	6	8	10	4	5	6	6	6	5	10	12
Great Britain	11	11	5	12	6	8	10	4	5	6	6	1	5	10	12
Croatia	7	10	5	12	6	8	10	4	5	6	6	6	5	10	12
Latvia	11	11	5	12	6	8	10	4	3	6	6	4	5	10	12
Slovenia	8	11	5	11	6	8	10	4	5	6	6	6	5	10	12
Estonia	11	11	5	9	6	8	10	4	5	6	6	6	5	10	12
Kazakhstan	11	9	5	12	6	8	10	4	5	6	6	6	5	10	12

Table 16. Examples of the results used in the second stage

Using the results found in the Table 16 we built a new adjacency matrix of the Condorcet Method and then we apply again the Copeland Method in order to establish a final ranking. In the next section the results will be shown and analyzed.

5. Results

After we used the method presented before, we have built a table by doing a paired comparison of the countries, taking into account the criteria established (in our case, each ranking of every modality). The result is a square matrix in which each country is compared with each other. The principal diagonal of the adjacency matrix is blank. In each other cell are computed the comparison between the country in the row and the candidate in the column. A score of 1 is signaling a pairwise victory for the country in the row over the country in the column. If happens a draw, both, row and columns of the pairwise in the matrix will get a score of 1.

Here we showed an excerpt of the adjacency matrix of the Condocert Method.

COUNTRY	United States	Germany	Canada	Norway	Austria	Russia	South Korea	China	Sweden	France	Switzerland	Netherlands	Czech Republic	Poland	Italy	Japan	Finland	Australia	Belarus	Slovakia	Croatia	Slovenia	Latvia	Great Britain	Kazakhstan	Estonia
United States		1	0	1	1	1	1	1	1	1	1	1	1	1	1	1	1	1	1	1	1	1	1	1	1	1
Germany	1		0	1	1	1	1	1	1	1	1	1	1	1	1	1	1	1	1	1	1	1	1	1	1	1
Canada	1	1		1	1	1	1	1	1	1	1	1	1	1	1	1	1	1	1	1	1	1	1	1	1	1
Norway	0	0	0		1	1	1	1	1	1	1	1	1	1	1	1	1	1	1	1	1	1	1	1	1	1
Austria	0	0	0	0		1	1	1	1	1	1	1	1	1	1	1	1	1	1	1	1	1	1	1	1	1
Russia	0	0	0	0	1		1	1	1	1	1	1	1	1	1	1	1	1	1	1	1	1	1	1	1	1
South Korea	0	0	0	0	0	0		0	0	0	0	1	1	1	0	1	1	1	1	1	1	1	1	1	1	1
China	0	0	0	0	0	0	1		1	1	1	1	1	1	1	1	1	1	1	1	1	1	1	1	1	1
Sweden	0	0	0	0	0	0	1	1		0	0	1	1	1	0	1	1	1	1	1	1	1	1	1	1	1
France	0	0	0	0	0	0	1	0	1		0	1	1	1	1	1	1	1	1	1	1	1	1	1	1	1
Switzerland	0	0	0	0	0	0	1	1	1	1		1	1	1	1	1	1	1	1	1	1	1	1	1	1	1
Netherlands	0	0	0	0	0	0	0	0	0	0	0		1	1	0	1	1	1	1	1	1	1	1	1	1	1
Czech Republic	0	0	0	0	0	0	0	0	0	0	0	1		1	1	1	1	1	1	1	1	1	1	1	1	1
Poland	0	0	0	0	0	0	0	0	0	0	0	1	1		0	1	1	1	1	1	1	1	1	1	1	1
Italy	0	0	0	0	0	0	1	0	1	1	0	1	1	1		1	1	1	1	1	1	1	1	1	1	1
Japan	0	0	0	0	0	0	0	0	0	0	0	0	0	1	0		0	1	1	1	1	1	1	1	1	1
Finland	0	0	0	0	0	0	1	0	0	0	0	1	1	0	0	1		1	1	1	1	1	1	1	1	1
Australia	0	0	0	0	0	0	0	0	0	0	0	1	0	0	0	1	1		1	1	1	1	1	1	1	1
Belarus	0	0	0	0	0	0	0	0	0	0	0	1	0	0	0	1	0	0		1	1	1	1	0	0	0
Slovakia	0	0	0	0	0	0	0	0	0	0	0	0	0	0	0	0	0	0	1		1	0	0	1	1	1
Croatia	0	0	0	0	0	0	0	0	0	0	0	1	0	0	0	1	0	1	0	1		1	1	1	1	1
Slovenia	0	0	0	0	0	0	0	0	0	0	0	1	0	0	0	1	0	1	0	1	1		0	1	1	1
Latvia	0	0	0	0	0	0	0	0	0	0	0	1	0	0	0	1	0	1	1	1	1	1		0	1	1
Great Britain	0	0	0	0	0	0	0	0	0	0	0	0	0	0	0	0	0	0	0	1	0	1	1		1	1
Kazakhstan	0	0	0	0	0	0	0	0	0	0	0	0	0	0	0	0	0	0	0	0	1	1	0	1		1
Estonia	0	0	0	0	0	0	0	0	0	0	0	0	0	0	0	0	0	0	0	0	1	0	1	0	1	

Table 17. Adjacency Matrix - Condocert Method.

To calculate the Copeland Ranking it is necessary to compare each cell with its diagonal opposite. For each country in the row we do a subtraction between the sum of its pairwise victories (number of "1" in the row) and the sum of its losses (number of \1" in the column). These results are computed and a Copeland Ranking is established.

In Table 18 we show the results of the method proposed in this work and others types of rankings.

COUNTRY	RANKING 1 – LEXICOGRAPHIC	RANKING 2 – TOTAL NUMBER OF MEDALS	RANKING 3 – PROPOSED COPELAND METHOD
Canada	1	3	1
United States	3	1	2
Germany	2	2	2
Norway	3	4	4
Austria	9	5	5
Russia	11	6	5
China	7	8	7
Switzerland	5	11	7
France	12	8	9
Italy	15	15	9
Sweden	7	8	11
South Korea	5	7	12
Czech Republic	12	13	12
Netherlands	9	12	14
Poland	15	13	14
Finland	24	15	14
Australia	12	18	17
Japan	20	15	18
Croatia	21	18	19
Latvia	23	23	19
Slovenia	21	18	21
Belarus	15	18	22
Slovakia	15	18	23
Great Britain	15	24	23
Kazakhstan	25	25	25
Estonia	25	26	25
Albania	26	27	27
Albania	26	27	27
Algeria	26	27	27
Andorra	26	27	27
Argentina	26	27	27
Armenia	26	27	27
Azerbaijan	26	27	27
Belgium	26	27	27
Bermuda	26	27	27
Bosnia & Herzegovina	26	27	27
Brazil	26	27	27
Bulgaria	26	27	27
Cayman Islands	26	27	27
Chile	26	27	27
Chinese Taipei	26	27	27

Colombia	26	27	27
Costa Rica	26	27	27
Cyprus	26	27	27
Denmark	26	27	27
North Korea	26	27	27
Ethiopia	26	27	27
Macedonia	26	27	27
Georgia	26	27	27
Ghana	26	27	27
Greece	26	27	27
Hong Kong	26	27	27
Hungary	26	27	27
Iceland	26	27	27
India	26	27	27
Iran	26	27	27
Ireland	26	27	27
Israel	26	27	27
Jamaica	26	27	27
Kenya	26	27	27
Kyrgyzstan	26	27	27
Lebanon	26	27	27
Liechtenstein	26	27	27
Lithuania	26	27	27
Mexico	26	27	27
Moldova	26	27	27
Monaco	26	27	27
Mongolia	26	27	27
Montenegro	26	27	27
Morocco	26	27	27
Nepal	26	27	27
New Zealand	26	27	27
Pakistan	26	27	27
Peru	26	27	27
Portugal	26	27	27
Serbia	26	27	27
Romania	26	27	27
San Marino	26	27	27
Senegal	26	27	27
South Africa	26	27	27
Spain	26	27	27
Tajikistan	26	27	27
Turkey	26	27	27
Ukraine	26	27	27
Uzbekistan	26	27	27

Table 18. Comparison between the ranking proposed and the others rankings.

Other analysis, in terms of the variation among the types of rankings, is showed. In the follow table we calculated the variation between the ranking proposed and the Lexigraphic Ranking.

COUNTRY	VARIATION $(X_{tk} - X_{tk-1})$	QUADRATIC VARIATION $(X_{tk} - X_{tk-1})^2$
Canada	0	0
United States	-1	1
Germany	0	0
Norway	1	1
Austria	-4	16
Russia	-6	36
China	0	0
Switzerland	2	4
France	-3	9
Italy	-6	36
Sweden	4	16
South Korea	7	49
Czech Republic	0	0
Netherlands	5	25
Poland	-1	1
Finland	-10	100
Australia	5	25
Japan	-2	4
Croatia	-2	4
Latvia	-4	16
Slovenia	0	0
Belarus	7	49
Slovakia	8	64
Great Britain	8	64
Kazakhstan	0	0
Estonia	0	0
Albania	1	1
Albania	1	1
Algeria	1	1
Andorra	1	1
Argentina	1	1
Armenia	1	1
Azerbaijan	1	1
Belgium	1	1
Bermuda	1	1
Bosnia & Herzegovina	1	1
Brazil	1	1
Bulgaria	1	1
Cayman Islands	1	1
Chile	1	1
Chinese Taipei	1	1
Colombia	1	1
Costa Rica	1	1
Cyprus	1	1
Denmark	1	1

North Korea	1	1
Ethiopia	1	1
Macedonia	1	1
Georgia	1	1
Ghana	1	1
Greece	1	1
Hong Kong	1	1
Hungary	1	1
Iceland	1	1
India	1	1
Iran	1	1
Ireland	1	1
Israel	1	1
Jamaica	1	1
Kenya	1	1
Kyrgyzstan	1	1
Lebanon	1	1
Liechtenstein	1	1
Lithuania	1	1
Mexico	1	1
Moldova	1	1
Monaco	1	1
Mongolia	1	1
Montenegro	1	1
Morocco	1	1
Nepal	1	1
New Zealand	1	1
Pakistan	1	1
Peru	1	1
Portugal	1	1
Serbia	1	1
Romania	1	1
San Marino	1	1
Senegal	1	1
South Africa	1	1
Spain	1	1
Tajikistan	1	1
Turkey	1	1
Ukraine	1	1
Uzbekistan	1	1

Tabela 19. Variation between the Proposed Ranking and the Lexicografic Ranking.

To organize the results we can separate them into four categories. The first one includes countries that have almost the same results in the ranking 1 and 2 but in ranking 3 they lost some positions. As examples are Sweden and South Korea. These results show that the medals won by these countries are balanced and they have invested in sports that distribute a large number of medals.

The second cluster is composed by countries that have the same (or almost the same) results in the ranking 1 and 2 and they win some positions in ranking 3. In this class we have countries as Italy and Latvia. In this situation are countries that have won a balanced number of gold, silver and bronze medals and also have invested in modalities that offer a small number of medals.

The third set of countries includes those with the same position in ranking 2 and 3 but in lower position in the ranking 1 as Austria and France. Some of them have the higher quadratic variation as Finland. These results indicate that probably the investments politics in sports of these countries are based in team-based sports which are not a good option to win medals. Another analysis is that they have won more silver and bronze medals than gold medals.

The fourth category is given by countries that have the same position (or a difference of one or two positions) in all the three rankings and the lowest quadratic variation, as Canada, United States, Germany, Norway, China, Czech Republic and Poland. These countries invested equally in a search for all types of medals in the modalities that distribute a good number of medals.

We can also take into account a case of countries as Switzerland. They have won a number of gold medals large than the others types of medals, but they invested in sports that allocate a small number of medals.

6. Conclusion

In this work we have proposed a study of the results of Vancouver 2010 Olympic Games. We have obtained some interesting results such as countries that have an unbalanced number of medals and take part in modalities that distribute a small number of medals.

Another interesting aspect found in this study is the Finland results. In the evaluation, we found that it probably invest in sports based in teams in which the number of medals are smaller than in other modalities and also it had won more silver and bronze medals than gold medals. This result corroborates the findings in Bergiante and Soares de Mello (2010) and helps us to understand some peculiar political sports decisions took by a couple of countries.

As a future work suggestion, an interesting proposal is to cluster countries by modalities in order to build others Olympic Rankings [Soares de Mello et al (2009)]. Another option is to explore the economic aspect of these conclusions and develop a model of investment politics in sports for each country.

7. Acknowledgments

We acknowledge the financial support of FAPERJ.

8. References

Allen, T.L., Jolley, S.J., Cooley, V.J., et al. (2006). The epidemiology of illness and injury at the alpine venues during the Salt Lake City 2002 Winter Olympic Games. Journal of Emergency Medicine, 30, 197-202 .

Arrow, K.J. (1951). Social choice and individual values. John Wiley & Sons, New York.

Ball, D. W. (1972). Olympic Games Competition: Structural Correlates of National Success. International Journal of Comparative Sociology 12,186-200.

Balmer, N. J., Nevill, A. M., Williams, M. (2001). Home advantage in the Winter Olympics (1908-1998). Journal of Sports Sciences, 19, 129-139.

Balmer, N.J., Nevill, A.M., Williams, A.M. (2003). Modeling home advantage in the Summer Olympic Games. Journal of Sports Sciences 21(6), 469-478.

Barba-Romero, S., Pomerol, J.C. (1997). Decisiones multicriterio: fundamentos teoricos e utilizacion prática. Madrid: Coleccíon de Economia, Universidad de Alcalá.

Bergiante, N., Soares de Mello, J.C.B. (2010). A DEA study of Vancouver 2010 Winter Olympic Games. Anais do XLII SBPO, Bento Gonçalves.

Bernard, A.B., Busse, M.R. (2004). Who wins the Olympic games: economic resources and medal totals. Rev. Econ. Stat., 86, 413-417.

Bernstein, E. (2000).Things You Can See from There You Can't See from Here: Globalization, Media, and the Olympics. Journal of Sport and Social Issues 24, 351-69.

Blain N., Boyle R., and O'Donnell H. (1993). Sport and National Identity in the European Media. .Leicester University, Leicester.

Boaventura Neto, P.O. (2003) Grafos: teoria, modelos, algoritmos. Edgard Blücher, São Paulo.

Cheng, X. (2009). The urban system impact on post-games development of the Olympics' venues in china. Paper read at 2009 International Association of Computer Science and Information Technology - Spring Conference, IACSIT-SC.

Churilov, L., Flitman, A. (2006). Towards fair ranking of Olympics achievements: The case of Sydney 2000. Computers and Operations Research. 33 (7), 2057-2082.

Farrell, T. (1989). Media Rhetoric as Social Drama: The Winter Olympics of 1984. Critical Studies in Mass Communication 6, 158-82.

Glynn, M. (2008). Configuring the field of play: how hosting the Olympic Games impacts civic community. Journal of Management Studies, 45 (6), 1117-1146.

Gomes, E.G., Soares de Mello, J.C.C.B.; Souza, G.S., Angulo Meza, L., Mangabeira, J.A.C. (2009). Efficiency and sustainability assessment for a group of farmers in the Brazilian Amazon. Annals of Operations Research, 169 (1), p 167-181.

Gomes Junior, S.F., Soares de Mello, J.C.C.B., Soares de Mello, M.H.C. (2008). Utilização do método de Copeland para avaliação dos pólos regionais do CEDERJ. Rio's International Journal on Sciences of Industrial and Systems Engineering and Management, 2 (1), p. 87-98.

Hadjichristodoulou, C., Mouchtouri, V., Vaitsi V., et al. (2006). Management of environmental health issues for the 2004 Athens Olympic Games: is enhanced integrated environmental health surveillance needed in every day routine operational. BMC Public Health, 6, 306.

Heazlewood, T. (2006). Prediction versus reality: The use of mathematical models to predict elite performance in swimming and athletics at the Olympic Games. Journal of sports science and Medicine,5,541-547.

Hilvoorde I., Elling A., Stokvis, R. (2010). How to influence national pride? The Olympic medal index as a unifying narrative. International Review for the Sociology of Sport; 45, 87.

Johnson, D.K.N., Ali, A. (2004).A Tale of Two Seasons: Participation and Medal Counts at the Summer and Winter Olympic Games. Social Science Quarterly , 85 (4).

Levine, N. (1974). Why Do Countries Win Olympic Medals? Some Structural Correlates of Olympic Games Success: 1972. Sociology and Social Research, 58, 353-60.

Li, Y., Liang, L., Chen, Y., Morita, H. (2008). Models for measuring and benchmarking Olympics achievements. Omega International Journal of Management Science. 36 (6), 933-940.

Lins, M.P.E., Gomes, E.G., Soares de Mello, J.C.C.B., Soares de Mello, A.J.R. (2003). Olympic ranking based on a zero sum gains DEA model. European Journal of Operational Research. 148 (2), 312-322.

Lozano, S., Villa, G., Guerrero, F., Cortfies, P. (2002). Measuring the performance of nations at the Summer Olympics using data envelopment analysis. Journal of the Operational Research Society. 53(5),50-511.

Roy, B. and Bouyssou, D.: Aide multicritere la decision: methods et cas. Economica, Paris (1993).

Roy, B. (1992).Decision science or decision aid science? European Journal of Operational Research 66, (2), p.184-203.

Soares de Mello, J.C.C.B., Angulo-Meza, L., Branco da Silva, B. P. (2009). A ranking for the Olympic Games with unitary input DEA models. IMA Journal Management Mathematics. 20, 201-211.

Soares de Mello, J.C.C.B., Angulo-Meza, L , Lacerda, F.G., Biondi Neto, L. (2009). Performance team evaluation in 2008 Beijing Olympic Games. In: XV International Conference on Industrial Engineering and Operations Management – ICIEOM.

Soares de Mello, J.C.C.B., Gomes, E.G., Angulo-Meza, L., Biondi Neto, L. (2008).Cross Evaluation using Weight Restrictions in Unitary Input DEA Models: Theoretical Aspects and Application to Olympic Games Ranking. WSEAS Transactions on Systems. 7 (1), 31-39.

Soares de Mello, M.H.C., Quintella, H.L.M.M., Soares de Mello, J.C.C.B. (2004). Avaliação do desempenho de alunos considerando classificações obtidas e opiniões dos docentes. Investigação Operacional, 24 (2),.187-196.

Streets, J.S et al. (2007). Air quality during the 2008 Beijing Olympic Games. Atmospheric Environment, 41, 480-492.

Wallechinsky, D. (2004). The Complete Book of the Summer Olympics. Aurum Press, New York.

Weiler,J.M. Layton, T., Hunt, M. (1998) Asthma in United States Olympic athletes who participated in the 1996 Summer Games. J. Allergy Clin Immunol,102 (5), 722-72.

Wu, J., Liang, L., Wu, D., Yang, F. (2008). Olympics ranking and benchmarking based on cross efficiency evaluation method and cluster analysis: The case of sydney 2000. International Journal of Enterprise Network Management, 2 (4), 377-392.

Wu, J., Liang, L., Chen, Y. (2009). Dea game cross-efficiency approach to Olympic rankings. Omega, 37 (4), 909-91.

Wu, J., Liang, L., Yang, F. (2009). Achievement and benchmarking of countries at the summer olympics using cross eficiency evaluation method. European Journal of Operational Research, 197 (2), 722-730.

Xiaoduo, C., Jianxin, Y. (2008). The factors of the urban system influenced postdevelopment of the Olympics' venues. Paper read at 2008 International Conference on Wireless Communications, Networking and Mobile Computing, WiCOM.

Yang, F., Ling, L., Gou, Q., Wu, H. (2009). Olympics performance evaluation and competition strategy based on data envelopment analysis. Paper read at Proceedings - 2009 International Conference on Computational Intelligence and Software Engineering, CiSE.

Zhang, D., Li, X., Meng, W., Liu, W. (2009). Measuring the performance of nations at the Olympic Games using DEA models with different preferences. *Journal of the Operational Research Society, 60 (7), 983-990.*

9

Quality of Life Modelling on the Basis of Qualitative and Quantitative Data

Jiří Křupka, Miloslava Kašparová, Jan Mandys and Pavel Jirava
University of Pardubice, Faculty of Economics and Administration
Czech Republic

1. Introduction

In the chapter the problematic of the Quality of Life (QL) modelling in the Czech Republic (CR) is presented. These models are suggested on the basis of the definitions QL analysis, approach to the measuring of QL and with using the complex system model of the QL, which can be used as a feedback information in the process of managing the public administration. At the basis of the systematic approach the models working with qualitative and quantitative data were suggested. In the first case it is the data from the questionnaire survey using the different methodology approaches. They are the data sets from the Institute of Sociology of the Academy of Sciences (ISAS) of the CR and a project of the civic association Team Initiative for Local Sustainable Development (TILSD). The methods cluster analysis (CA) and decision trees (DTs) were used for classifying the "satisfaction of respondents" with the selected part of QL and recommended decision making regulations. In the second case it was the quantitative data from the public accessible sources in the CR, e.g. Public database of the Czech Statistical Office (CSO), Institute of Health Information and Statistics of the CR, Czech Hydrometeorological Institute, CSO of Pardubice's region, Portal of Regional Information Service etc. When selecting the indicators of QL, which represent the input variables of the model, it was resulted of the Strategic Framework (SF) of the sustainable development in the CR. The methods regression analysis were used for predicting the values of the indicators, DTs for classification of the QL levels in the regions of the CR and methods of the multidimensional statistics for the suggestion of the aggregated indicators of QL.

An expression QL is closely related to an expression life's satisfaction. It is obvious that defining of the expressions satisfaction and QL brings many dilemmas and it is necessary to take into considerations the differences in the opinion of this theme. Reaching the life's satisfaction is either conscious or unconscious effort of every individual. It is very subjective figure, which is changing in time and the expression "satisfaction" itself is vast and indefinite. Also the satisfaction determines, to the considerable extent, individual perception of the QL. This expression can be partly overlapped in some theories with well-being and life satisfaction could be a superior expression (Baštecká & Domkařová, 2011).

An asset to the creation of the opinion "satisfaction" is presented in the work (Hamplová, 2006), who summarizes pieces of knowledge of many authors, such as (Diener & Lucas,

2000; Kim & Hatfield, 2004; Stach & Eshleman, 1998, as cited in Hamplová, 2006), who characterize this expression as a state, when the positive emotions predominate the negative and in satisfaction then it is reflected a sensible evaluation of the own life. With reference to (Averill & More, 2000, as cited in Hamplová, 2006), satisfaction (the expression correlates with the expression happiness) is the state of joy, high spirits, peace and poise. In this case on the other hand, the expression happiness brings a lot of other questions, not really closer specification to the overall problem.

Much more pragmatic is the opinion of satisfaction (Půček et al., 2005). He uses the common reactions of the human who answer the question "What does this expression mean to you?" The human usually answer either "When I do well" or "When everything goes according to my wishes."

In reality it means the attitude (perception) of the specific state or situation, to what extent the human fulfilled his/her expectations of need, wishes etc. Satisfaction reflects the degree of the requirements fulfilment. This perception is then connected to the emotions and presents very subjective (relative category).

If we deal with the defining the expression QL, we must realize the historic, social and cultural changes in the society. Moreover by (Duffková et al., 2008), the QL includes the individual way of life, life conditions of the individuals, groups or society as a whole.

A great number of the literature brings different views of QL e.g. (Dvořáková et al., 2006; Philips, 2006; Rapley, 2003; Tokárová, 2002a; Tokárová et al., 2005) and also it argues that the QL is greatly discussed theme for many disciplines. As it is described by (Řehulková & Řehulka, 2008; Tokárová, 2002b) the QL was at first the economy object of interest (20s in the 20th century) as an indicator of satisfaction and social welfare, which is not primarily influenced by the number of consumed goods, but is presented as a subjective experience perception. Then the expression spread into sociology (50s in the 20th century) and psychology and medicine (in this case the QL represents a very important aim). Interdisciplinary approach towards the overall phenomenon brings a lot of advantages. The subjective as well as objective aspect can be expressed. On the other hand this complexity of the overall phenomenon does not enable to create a unified generally applicative model, but the result approach is always influenced by the professional interest of the researcher of the theories, which he/she prefers. This can cause often quarrels during the discussions and the results which can be applicable in reality, e.g. when strategic decision making of the public administration, must be interpreted with paying attention to all possible aspects of the solved reality. Difficult determinableness of the expression is in the subjectivity, when every individual understands it differently. By (Křivohlavý, 2004, p. 10) "About QL it is possible to discuss and have in mind a different range of an expression - human".

Complexity of the phenomenon also shows the characteristics and determinants of the QL where all the fields of the human life are reflected. Besides the factors which are mentioned below, we must realize (Blažej, 2005), that the 21st century brings new tendencies and aspects in the QL; from the most important ones are: globalization, development of the information activities, sustainable development and new economy influenced by the actual pieces of knowledge, where the important influence has the human capital to the socio-economical development. World Health Organization (WHO) differentiates four basic dimensions of the human life determining its quality. They are absolutely independent on the factors of age, gender, ethnic or disability (Műhlpachr, 2005, p. 61):

- Physical health and level of independency – energy and tiredness, pain, relaxation, mobility, everyday life, dependence on the medical help, ability to work etc.
- Mental health and psyche – self-conception, negative and positive feelings, self-evaluation, thinking, learning, memory, concentration, belief, spirituality, religion etc.
- Social relationships – personal relationships, social support, sexual activity etc.
- Environment – financial resources, freedom, safety, accessibility of health and social care, domestic environment, opportunities for gaining new knowledge and skills, physical environment (pollution, noise, traffic, climate) etc.

The social determinants of the QL are thought these (Halečka, 2002, p. 67):

- Complex, optimal environment
- Adequate usage of human activity and drive
- Overall quality of human relationships
- Developing division of the competences and conceptual routing of the other human development
- Full respect of the human dignity as a bio psychosocial personality
- Mutual contribution to the higher values realization, to the full human being, his/her transcendence

Synthesized social indicators (conditions of life) in their work are presented by (Vaďurová & Műhlpachr, 2005, p. 8). These are[1] :

- Health
- Quality of working environment
- Purchase of the goods and services
- Possibility of free time spending
- Feeling of the social certainty
- Possibilities of the personality development
- Physical quality of the environment
- Possibility of social life attendance

2. Problem formulation

The way how to study the QL (Fig. 1) is offered by (Blažej, 2005, p. 25), who also encourages an acceptance of the social and international objectively measurable criteria, which would be studied and evaluated regularly. Considering different approaches determining the QL seems the creation of generally accepted categories unreal, as the other text shows it is very hard to express all the aspects of human life in context of his/her subjective and objective sides. Another question is how much this synthesis is necessary and covetable, as it is highlighted e.g. by (Miovský, 2006) scientific cognition is not united, homogenous complex of methods or universal rules, by which this cognition should be realized.

---

[1] Synthesis was realized in 1974 by the European Commission of the United Nations

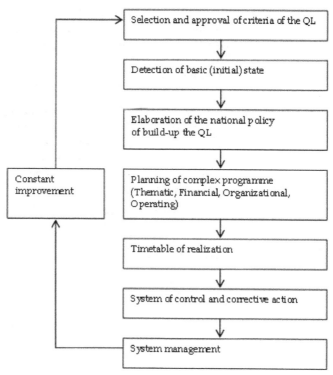

Fig. 1. Complex system of the QL.

The QL is necessary to view as a subjective appraisal of the own life situation. The QL consists of data about psychosocial state of the individual, which are influenced by the factors such as age, gender, education, social status, economic situation, values or personal well-being (Philips, 2006). Similarly the QL is defined by (Vaďurová & Műhlpachra, 2005), who say that this expression means the individual's perception in the place of life. This perception is influenced by culture, value system, by relations towards the human's aims, expectations, norms and worries. As the other variables there is also psychosomatic condition of the individual, social relations, personal religion and also the relation towards the key fields of his/her environment.

If we deal with the analysis of approaches towards the QL, we can study five main directions; which concepts influence the most understanding the QL, compare e.g. (Dvořáková et al., 2006; Křivohlavý, 2011; Maříková et al., 1996; Možný, 2002; Payne et al., 2005; Phillips, 2006; Vaďurová & Műhlpachr, 2005):

- Psychological concept (or socio-psychological): Psychology is in the given example oriented to the individual aspects of experiencing feelings of personal satisfaction and well-being. Sometimes the expression happiness is used. Foreign terminology related to the QL also uses besides the expression of well-being, the expression QL, subjective or psychological, life satisfaction, mental health status or happiness. Well-being then de facto means the long-lasting emotional status, marks out by the time stability and consistence in different situations. Personal well-being of the individual is leant on the cognitive components, such as life's satisfaction or moral principles. In more details about construct of well-being (Šolcová & Kebza, 2004)

- Sociological concept: According to the sociological view there are also other factors (social) besides the individual factors (e.g. culture, religion, health, income, age, job satisfaction, mobility, transport etc.). The object of the interest is then the standard of living, way of life or welfare, which is represented as welfare in the foreign literature
- Philosophical concept: Behind the important characteristics of the QL we can consider meaningfulness of life (heading towards). If the human sets the superior general aim of life, then the aim is the main indicator of his/her life's meaningfulness. This aim shows where the sense is and where not. The big role plays the conscience, which must balance the procedures chosen when filling the aim to reach it. The expression of QL is related to understanding to the human existence
- Physiological concept (medical): A big amount of literature deals with the problematic of the QL at the people with illnesses. And in the field of measuring the QL it can be very well connected to the problematic of subjective and objective perception of the QL. Definition of the QL can help us, which is set by WHO, where the health is understood as a state of total physical, mental and social well-being. This definition takes note of psychologically-sociological aspects as well as biological aspects in the expression of health. Value of health is generally accepted throughout the human cultures
- Economic concept: This concept goes from mainly objective indicators of the QL (method of measuring, which will be described below). Because the QL from this point of view goes from the consumption, wealth, poverty and globalization, in our conception it is rather logical framework of the overall problem

All advanced views are partly taken in the work by (Rapley, 2003), who presents the options of the QL research according to these areas:

- QL as a psychological object
- QL of the selected population groups
- QL from the view of health and social care
- QL as an object of cultural impact

As one of lots of models of the QL can be illustratively used e.g. graphical visualization of the model (Rapley, 2003, p. 54) including essential factors of the personal well-being, which also contains all important variables (objective or subjective), which creates the QL (Fig. 2). That the QL is an object of a high interest is supported by (Vaďurová & Mühlpachr, 2005, pp. 9-10) list of organizations and research groups (Table 1).

The QL can be considered as universal category, which describes the satisfaction of people in their personal as well as social life. This information can then be useful for creation of local developing activities at the certain area. It is necessary to choose the adequate method of the collection of this data (for their usage) in respect to multidimensionality of the phenomenon.

In the basis, we can divide the approaches of the research into two main groups and one associated. For indicating the actual reality it seems to be the best to use both approaches together. Due to the demandingness of the research this solution seems to be hard realizable for practical usage, because it collides with the options of the research submitter and his/her

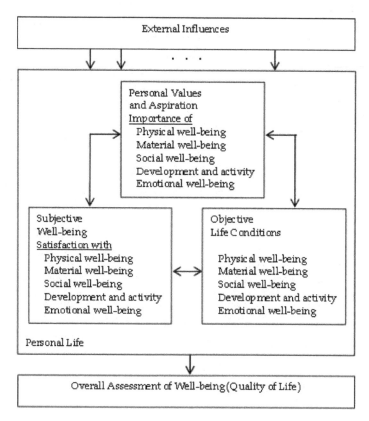

Fig. 2. Model of the QL.

willingness to invest into this process of data collection. The main parameter of the approach of the QL research is also reviewing the different indexes and parts of the values of respondents by variable scales:

- Subjective indicators of the QL: we can say simply that the personal satisfaction is measured, opinions of the people towards the certain phenomenon. Among the methods we can de facto include every researching technique, which ask the respondents their satisfaction. (Vaďurová & Mühlpachr, 2005) say, that the evaluator is the person himself/herself. In more details we will deal with the particular tools below

- Objective indicators of the QL: into this category we can include internationally accepted indicators of the state of the society and its development (see below). (Vaďurová & Mühlpachr, 2005) say, that in this case the objective procedure of the measurement can be considered the situation, when the state (situation) is evaluated by the second person

- Combined indicators: are represented by collected constructions of both above

Organization and Research Groups	Organization and Research Groups
American Thoracic Society – QOL Group	EORTC Quality of Life Study Group
Australian Centre on Quality of Life	Health Services Research – UCLA
Australian Health Outcomes Collaboration	Health Assessment Lab (HAL)
Behavioural Sciences at Nottingham University	Health & Quality of life Research Centre
Cancer and Leukemia Group B (CALGB)	Health & Survey Research Unit
Cardiff Research Consortium	Health Outcomes Research Europe
Center for Health Outcomes, Policy and Evaluation Studies (HOPES), The Ohio State University	National Centre for health Outcomes Development (include the patient-assesssed Health Instrument database)
Centers for Disease Control and Prevention's Division of Adults and community Health	Health Services Research Unit, University of Oxford
Center for Health Program Evaluation (CHPE)	Health Utilities Group (HUG)
Center for Health Outcomes and Policy Research, University of Pennsylvania	Human Research Services Institute
Center for Health Quality, Outcomes § Economic Research (CHQOER).	Institute for Health Services Research and Policy Studies
Center for Outcomes Research, University of Massachusetts	International Society for Pharmacoeconomics and Outcomes Research (ISPOR)
Center for Pharmaceutical Outcomes Research, University of North Carolina at Chapell Hill Scholl	International Society for QOL Studies (ISQOLS)
Chartered Society of Physiotherapy	International Society for QOL Research (ISQOL)
Cochrane Collaboration	Irish Clearing House on Health Outcomes
Connaissances et Décision en Economie de la Santé (base CODECS)	MAPI Research Institute
Department of Health Care Policy Harvard Medical School	Medical Outcomes Trust
Department of Health Services, University of Washington	Medical technology & Practice pattern institute (MTPPI)
Department of medicine, Clinical Epidemiology and Biostatics, McMaster University	European Clearing Houses on Health Outcomes (ECHHO)
Department of Palliative Care and Policy, London	

Table 1. Main organization and research groups dealing with the QL.

Besides the translated typology of division, we can distinguish the methods according to their specifications.

2.1 Methods of survey of subjective and objective QL

It is necessary to remind, that subjective indicators can be considered all answers of the human about his/her situation, which reflects his/her attitudes. Among the methods of survey of the subjective QL belongs: questionnaire of the WHO QL (WHOQOL) and its equivalents, Lancashire QL Profile (LQoLP), Schedule for the Evaluation of Individual QL (SEIQoL), questionnaire SQUALA and others.

In case of questionnaire WHOQOL and other equivalents it is one of the most used methods of studying the QL. Authors of questionnaire WHOQOL-100 conclude from the definition of the QL, which says that the QL is how the individual perceives his/her position in life (in the cultural context, in relation to his/her aims, expectations and interests). WHOQOL-100 consists of 24 aspects of life unified into 6 domains, where there are e.g. physical health, experiencing, level of independence, social relations, environment, spirituality and overall QL. The questionnaire is intended for population up to 65 years old. For older people the modification WHOQOL-OLD is used. WHOQOL-100 distinguishes among the groups of people with different level of health problems and between the men and women. Retested reliability of questionnaire domains WHOQOL-100 measured in the range of two weeks shows the relative stability of the answers in this time interval. An alternative can be a usage of the questionnaire WHOQOL-BREF. It consists of 24 items unified in 4 domains and two separate items evaluating an overall QL and health condition (26 items altogether). The questionnaire is not suitable to use for registering the impact of the immediate mood or the short-term changes (Miovská, 2011; Dragomirecká, 2006). For the people with HIV then the specific questionnaire is used WHOQOL-HIV and WHOQOL-HIV-BREF (Vaďurová & Műhlpachr, 2005).This questionnaire and its variations is primary used for measuring the health problematic and its outputs can be used also for the results in the ordinary population.

A questionnaire LQoLP is the questionnaire combining the subjective and objective sides of the QL. It consists of the following fields: work and education; free time; religion; finance; life situation; law status and safety; relationships with the family; social relations and health. The questionnaire is used for work with people who are mentally ill (Oliver, 1992; Vaďurová & Műhlpachr, 2005).

Other very much used method of the survey the quality is presented by (Křivohlavý, 2011; O'Boyle et al., 1993; Vaďurová & Műhlpachr, 2005) SEIQoL. It is the basic research technique, which is a structured interview. Interviewer finds out five basic life aims (impetus for life) of the respondent. Respondent indicates the level of satisfaction at all aims. The most often the impetus health, family, work, religion, finance, education, culture, hobbies etc. are mentioned.

A questionnaire SQUALA (Dragomirecká et al., 2006; Vaďurová & Műhlpachr, 2005) goes from Malowov theory of needs. It concludes from the opinion, that the QL means finding out the difference between the wish and expectations of the individual and the real state how this wish is fulfilled. The questionnaire is divided into two parts of the satisfaction evaluation. The evaluation of importance means: be healthy; be physically independent; feel mentally fit; nice environment and living; sleep well; family relationships; relationships with other people; have and educate children; take care of ourselves; love and be loved; have a sexual life; be interested in politics; believe in something (e.g. religion); relax in the free time; have hobbies in the free time; be in safety; work; justice; freedom; beauty; truth; money and good food. To the evaluation of satisfaction it is included: health, physical independency, mental well-being, environment of living; sleep; family relationships; relations with other people; children; care of ourselves; love; sexual life; membership in politics; belief; relax; hobbies; feeling of safety; justice; freedom; beauty and art; truth; money and food.

Among the other methods of survey the QL we can include e.g. (Vaďurová & Műhlpachr, 2005): Behaviour and Symptom Identification Scale (BASIC 32), Groningen Social Disabilities Scale (GSDS-II), General Satisfaction Questionnaire (GSQ), Psychosocial Adjustment to Illness Scale (PAIS), QL Enjoyment and Satisfaction Questionnaire (Q-LES), Social Behaviour Schedule (SBS) and Social Functioning Schedule (SFS), and others.

Among the methods of survey of the objective standards of living are: Human Development Index (HDI), Gross Domestic Product (GDP), Standard of Living of the Households or Individuals (SLHI) etc.

HDI consists of these three components: wealth, health and education. At these items were set minimal and maximal fixed values: 25 and 85 years (average length of life – hope of spending the rest of life of the member of the population); 0 and 100 % (literacy of population older than 15 – school attendances, length of studies etc.); 0 and 100 % (combined portion of population from the age group attending schools of first, second and third grade); 100 and 40,000 USD (GDP per person in parity of purchasing power). This index was topped up with the index of gender, index of women presence in social life (Blažej, 2005; Charles University Environment Centre, 2010; Tokárová, 2002b).

GDP presents the financial expression of the overall value of the estates and services newly created in the given period of time at the specific area. Material welfare, which is expressed by the GDP, financial expression of the overall estates and services newly created in the given period of time at the specific area. It is about the standard of living of the society (Kubátová, 2010; Czech Statistical Office [CSO], 2010). It is necessary to view this indicator critically, because (Možný, 2002, p. 17):

- Assumption of that people always select things according to their benefits and towards their benefits is not always true. People do not often select what they need
- Maximization of the individual benefits leads to the maximization of the social welfare
- Our behaviour does not bring any intended consequences to the others, and if yes, they can be ignored
- Distribution of the income (or division of the prosperity) is technically all right
- Those, who make a choice (spend money), are always those, who consume things or have some benefits from them
- All kinds of consumption are equal
- Things have the value connected to the market value right now

SLHI means the standard of income and consumption. They give evidence about the wealth and poverty. Particularly it presents the direct numeration of consumed goods and services, or pertinently financial income and estates, free time, resources from the budget paid for the public services. Then there is also number of harmful substances discharging to the water or polluting the air, average life expectancy, infant mortality, crime rates (i.e. demographic indicators). In other words expressing the standard of living consider, that we can imagine it as a degree of answering the material and non-material needs, wishes of the individual or the household. It is expressed by goods and services in particular it means the relation between the real state and the wished (satisfying) state (Červenka, 2011; Kubátová, 2010).

All these indexes can be understood with some reserve in their given evidence value. On the other hand they are the internationally accepted indicators, which are able to be compared, and they present the guidance (practical) data studied throughout the cultural specifications of the individual nations (Možný, 2002).

A lot of methods of the research of the QL are concentrated to the medicine. For example (Křivohlavý, 2011; Vaďurová & Mühlpachr, 2005): Index of the quality of patient's life (ILF), Acute Physiological and Chronic Health Evaluation System (APACHE II) and others. The core of the method ILF is that the subjective evaluation of the patient is added by the opinion of the other interested people (medical personnel etc.). Among the studied criteria belong, e.g. patient's self-service, social support of the patient, managing the distress connected to the illness, pain of the patient and overall emotional state of the patient etc. APACHE II is the evaluating system of the acute and chronically changed health and gives a true picture of the overall state of the patient by means of physiological and pathophysiological criteria. The other methods are: Karnofsky index (The Karnofsky Performance Scale), Visual Analougue Scale (VAS), Spitzer Duality of Life Index – QL etc.

The problem of the QL in the CR is one of the main parts of the Strategy of national policy of quality in the CR for the period 2008 to 2013. The concept of the strategy (Vorlíček, 2008) is based on a result analysis of the present fulfilment of the national policy of quality support and on the basis of the evaluation of the current situation; it defines sending, vision, framework and the long-term strategic goal for the next period. The effort is to create an environment that would improve society life in all areas (including the improvement of the QL of individuals) in the CR (Křupka et al., 2009, 2011a).

Very important external factors of the QL are different regulatory instruments. These can be divided into several categories. First, they are laws and regulations in the areas related to the QL and affecting the QL. It is the external influence which affecs many objective life conditions (see Fig. 2). Examples are strict standards of environmental protection and emission limits. These are forcing manufacturers to reduce emissions or cancel too "dirty" industrial plants. As a result, can this improve the quality of the environment that is important indicator of the QL (see the introduction of this chapter). Another instrument is municipal legislation. Municipal legislation states in general terms the jurisdiction of council and provides the legal structure and framework for municipal councils to provide governance and to make decisions at a local level. It affects the QL at the regional and local level. Finally, the regulatory instruments are subordinate standards and documents. Various institutions, associations, communities can issue such documents governing various aspects of the QL. People are governed by them voluntarily and on the basis of their beliefs. Their influence on the QL is not only local, only rarely broader.

The latest document within the CR is the SF of the sustainable development in the CR from 2010. It is a strategic material which is used as a long lasting guideline for a political decision making in the context of international obligations. The SF introduces four so called global aims according to the renascent Strategy of the sustainable development of the EU from 2006. They are these: Protection of the Environment, Social Cohesion, Economic; Prosperity and International Responsibility. Within these documents the sustainable development is defined as a development, which will carry out the needs of the present generation without threatening the needs of the future generations. There are priority axis and aims defined because of reaching a desirable situation of the sustainable development. According to (Ministry of Environment CR, 2010) they are these: Priority axis 1 – Society, Human, Health; Priority axis 2 – Economics and Innovation; Priority axis 3 – Area Development; Priority axis

4 – Countryside, Ecosystems and Biodiversity; Priority axis 5 – Stable and Safe Society (Křupka et al., 2011b, 2011c).

3. Quality of life modelling

When analyzing QL problems we work with notions as: quantitative and qualitative data, methods, research, evaluation and similar. Quantitative data (we can also use the terms "hard" data) are numeric characteristics (variables) of the observed phenomenon. Qualitative data ("soft" data) are non-numeric characteristics of the observed phenomenon (Křupka et al., 2011b). In quantitative research (Disman, 2005) multidimensional social and human reality is reduced to a limited number of a couple of variables and to a small number of analyzed relations between such variables. Qualitative research is a non-numerical examination and interpretation of social reality the objective of which is to uncover the meaning of the interpreted information.

On the basis of data sets of research inputs and output indicators (attributes, variables) for a model creation were defined.

3.1 Models work with qualitative data

These models are devoted to the issue of the design of "citizen satisfaction" classification models. Real opinions on the QL problem were taken in to consideration when creating the classification models "citizen satisfaction" with regards to the quality of environment at the regional level in the CR (Křupka et al., 2011a). We worked with two methodology approaches (ISAS and TILSD), we compared them and defined common categories, e. g. environment, education, heath service and possibility to participation in local decision making (Křupka et al., 2011a). The ISAS realizes monthly research in a wide spectrum (political, economical and social topics, respectively) covering the opinions of the inhabitants of the CR. The TILSD engages in the QL from the European Common Indicators (ECIs) point of view. The TILSD realizes a questionnaire inquiry for a given town, whereas the ISAS has a wide target group of informants and the place of residence represents only one of the variables (attributes) in the questionnaire inquiry. In regards to the representation of the QL, the TILSD uses an evaluation index of satisfaction and the ISAS uses a percentage expression (Křupka et al., 2011a).

On the basis of (Turban et al., 2005, p. 24) a model of citizen satisfaction might be defined in Fig. 3. The classification model of satisfaction was designed by means of algorithms of DTs (Maimon, & Rokach, 2005; Rokach, & Maimon, 2008) and their optimization. Based on the research conducted by the ISAS and TILSD in 2007 (Křupka et al., 2011a) real data sets were defined. In the first case (using the ISAS methodology) we worked with data matrix that included 1,132 results from a survey (objects) and 28 questions (attributes) from the questionnaire were selected. The modelling was focused on satisfaction of inhabitants with the environment in their place of residence using the ISAS methodology. Types of attributes are ordinal, nominal and continuous. The used data matrix does not content missing values and outliers. In the second case (using the TILSD methodology) we dealt with QL, it means citizens satisfaction with the quality of the environment in Chrudim. We used the data set from the questionnaire survey. On the basis of the chosen attributes come from the ECI indicator A1 "Citizen satisfaction with the local community", a data file was created, containing 701 data logging (objects, informants) and 10 original attributes (variables, questionnaire questions). Processes of data analysis, data preparation (data cleaning, data discretization and creation of binary attributes), splitting of data matrix, synthesis and analysis models, on the basis of decision trees (C5.0 and C5.0-boosting method), Classification and Regression Trees (C&RT), Chi-square Automatic Interaction Detection (CHAID), and Quick Unbiaset Efficient Statistical Tree (QUEST)), were realized.

The designed classification models were created in Clementine ver. 10.0. Standard type nodes were used in the models. Common methods were used in the evaluation of achieved results, see more in (Křupka et al., 2010a, 2011a).

The accuracy rate A_c belongs to easy characteristics determinative quality of found knowledge. General calculation of accuracy rate A_c is defined as follows:

$$A_c = (TP + TN) / (TP + TN + FP + FN), \qquad (1)$$

where TP is true positive, TN is true negative, FP is false positive and FN is false negative classification.

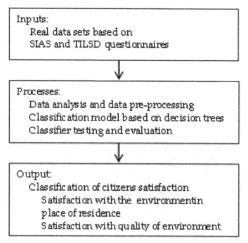

Fig. 3. Classification model of citizen satisfaction.

Partial results of classification accurasy A_c for model (it means model component works only with TILSD data) were compared with A_c of models (Křupka et al., 2010a) based on Radial Basis Function and Probabilistic neural networks (PNNs) , too (Table 2).

Methods of classification	A_c in %
C5.0	70.00
C5.0-boosting	70.00
CHAID	68.18
C&RT	67.73
PNN	66.52

Table 2. Average values of accuracy rate of selested classification models for TILSD data set.

3.2 Models work with quantitative data

These models are dedicated to the issue of the design of "quality of health state" classification models, regression models of a sustainable indicator and "composed indicator set" (see the chapter 3.2.1) on the basis of principal components analysis (PCA) and factor analysis (FA).

We focused on Quality of Health State (QHS) in the CR regions with using health and environmental areas. Designed QHS models (Fig. 4) use data received from public sources. These are real data coming from the Czech Statistical Office, the Institute of Health Information and Statistics of the CR etc. (Křupka, 2010b). It represents real yearly data for QHS from 1997 to 2007 from the regions of the CR and model defines three linguistic levels of QHS in the regions. QHS model works in two steps (Křupka, 2010b). In the first step a cluster analysis (CA) is used for linguistic level. Selected health attributes are inputs for CA. In the second step we apply selected algorithms of DTs (Maimon, & Rokach, 2005; Rokach, & Maimon, 2008) for classification model creation. Input matrix contains demographic attributes oriented to health, appendix attributes (number of physicians and hospitals) and selected environmental attributes. Used inputs have a wider significance and mutual relations, which are described below. The chosen attributes (parameters, indicators) are also selected in accordance with the probability of causes of death. Data file contains 140 objects (regions in the studied years) and 17 attributes.

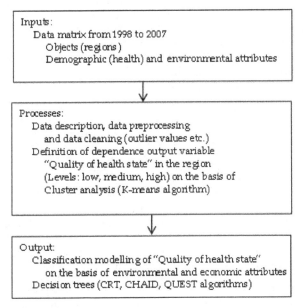

Fig. 4. Classification model of QHS.

When model creation, it is always very important to have a phase of analysis and understand the data. For the following analysis it is possible to use regression analysis. We suppose that there is an equation between the dependent y and independent x variables (Albright et al., 2006). The data is necessary to transfer to the relative values. We dealt with the analysis of selected, open accessible regional quantitative indicators from the priority axis 1 "Society, Human, Health" and 5 "Stable and Safe Society"of the SF (Křupka, 2011b). Priority axis 1 aims to develop and improve conditions for healthy lifestyle, improve the lifestyle and health of the population and adapt the state and the regional policy of the demographic development. The other topic of the axis is a family and an inter-generation cohesion. Priority axis 5 is aimed on the strengthening of a social stability in the society, on the development of an effective public administration and the state, the development of a civic sector and on handling the global, terrorist and other threats.

We presented basic indicators and super-structural indicators for the area "Stable, safe society, health and a human being" went through a simple consultation process. One of them is basic indicator "Number of offences in the region, from them disclosed cases".

This indicator shows the crime rate of the area. We have to realize that it brings the risk. This indicator works only with identified offences. None of the institution is not able to estimate, how much of real criminal (pathological) behaviour is in the society. However, we can consider this indicator as important indicator of the state of the society. In order to that we have to take all evidence with reserve and use other corroborative materials e.g. local documents about the crime prevention or use counsels of interested institutions (workers in the prevention, mainly in the authorities, social curators, non-governmental and non-profitable organizations, parts of the City Police or State Police etc.). The Indicator is possible to discuss only from the view of really pathological (deviant) behaviour compared to statistics of the registered offences. From the safety point of view, we can say that many deviations from the norms may cause some problems to people in their private or public life, but also they do not relate to criminal activity and therefore we cannot find any evidence. In many cases it is not possible to disclose them despite the fact that the crime happens. It is necessary to take into account this statistic evidence with the knowledge that there is not any other statistic evidence without factual findings (Křupka, 2011b).

The indicator is a basic indicator of disclosed offences in the years 2005-2009 for each district. For the years 1994-2004 the data is not possible to find for each district, but only for the separate regions. This is the crime introduced in the statistics of the Police of the CR – i.e. number of acts which are taken as offences – see the Act No. 140/1961Sb (the Criminal Law) and Act No. 141/1961Sb about prosecution (Criminal Rules).

We work with real data of districts (PCE is Pardubice, CHR is Chrudim, SVIT is Svitavy and UO is Ústí nad Orlicí) in Pardubice region for the years 2001-2009. We express these absolute values per 1,000 inhabitants. The equation of the regression (linear y_L and quadratic y_Q output of the regression model) is:

$$y_L = a_0 + a_1x \quad \text{and} \quad y_Q = b_0 + b_1x + b_2x^2, \tag{2}$$

where y_L and y_Q are expected values of offences in the territory; x is observed value of offences in the territory per 1000 inhabitants; a_0 and b_0 are constants; a_1, b_1, b_2 are coefficients of the model.

The coefficient of determination R-square R^2 (Albright et al., 2006; Ragsdale, 2008) was used for comparison of regression model quality. Model values for distrists of region are in the Table 3. Based on comparison of R^2 we can use linear trend of the district PCE, maybe for district CHR. We can not use linear trend for district SVIT and district UO like relevant information into regional planning process.

Territory	Value		
	a₀	a₁	R²
District PCE	31.881	- 1.022	0.7439
District CHR	19.244	- 0.334	0.5502
District SVIT	17.344	0.039	0.0153
District UO	16.670	- 0.022	0.0164

Table 3. Values of linear regression model for district of Pardubice region.

We defined quadratic regression model for Pardubice region, its R^2 is better than linear model, see Table 4.

Value of linear regression model			Value of quadratic regression model			
a₀	a₁	R²	b₀	b₁	b₂	R²
22.143	- 0.385	0.664	21.476	- 0.021	-0.036	0.695

Table 4. Values of linear and quadratic regression model for Pardubice region

The curve of quadratic regression model and observed values are in Fig. 5.

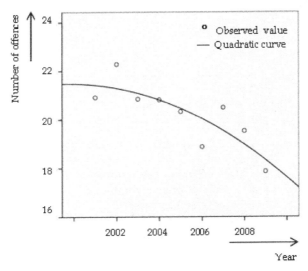

Fig. 5. Quadratic function of number of offences in Pardubice region.

3.2.1 Quality of life modelling by PCA and factor analysis

For modelling the data matrix formed by the districts of the CR studied in the years 2001, 2002, ..., 2008 was gained. After the phase of the analysis and clearing the data the resulting data matrix was used. It had the dimensions 42x529, i.e. we worked with 42 indicators (variables) p_i and 529 records (districts in the studied years). An example of the selected indicators is the following: area of the districts altogether p_1; density of the inhabitants p_2; average age of the inhabitants p_3; portion of the selected kinds of property – farmland from the overall area of the districts p_4; completed flats, room or the set of rooms that can be used as a single housing unit where there the occupation permit came into force in the studied period p_5; collective accommodation establishment p_6; average monthly income p_7; ...; made investments for the environment protection according to the districts of the investor's headquarters, single buildings and other investment measures leading to the improvement of the current environment conditions p_{10} etc. Detail description is in (Augustinová, 2010).

With intention of reducing the gained indicators, we focused on powerful techniques to reduce the complexity of data. Two similar but distinct approaches are used (SPSS Inc., 2007): There are PCA and FA. The PCA finds linear combinations of the input fields that do the best job of capturing the variance in the entire set of fields, where the components are orthogonal (perpendicular) to each other. PCA focuses on all variance, including both shared and unique variance. The FA attempts to identify underlying concepts, or factors, that explain the pattern of correlations within a set of observed fields. The FA focuses on shared variance only. Variance that is unique to specific fields is not considered in estimating the model. Principal components method and principal axis factoring method we used as methods for data reduction. The first one of methods uses PCA to find components that summarize the input fields. The second one is FA. It is very similar to the principal components method, except that it focuses on shared variance only (SPSS Inc., 2007). Design of modelling is in the Fig. 6.

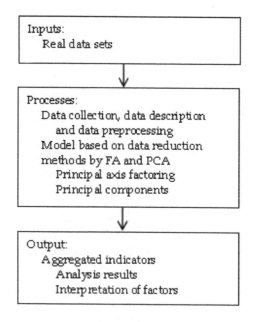

Fig. 6. Model of reduction indicators.

For the modelling the product of the company SPSS Clementine 12.0 was used and the above mentioned methods of variables reduction were applied. For the model creation the rotation method Varimax was selected (SPSS Inc., 2007): It is an orthogonal rotation method that minimizes the number of fields with high loadings on each factor. It simplifies the interpretation of the factors.

Factors were extracted using eigenvalues, output factors had the value of the eigenvalues greater than 1. Results of modelling we can see in the Table 5.

Model	Method	Number of Factors	Variance [in %]	Number of iteration for rotation
M1	Principal axis factoring	9	73.272	14
M2	Principal components	9	73.272	16

Table 5. Results of modelling by selected methods for data reduction.

Applying the mentioned methods of the variables reduction, the same number of factors was gained which gives a true picture of 73.3 % of the input data variance and the variables useful when single factors interpretation were identified. They were these factors: f_1 Health, f_2 Leisure activities, f_3 Social situation, f_4 Well-being, f_5 Density of the inhabitants, f_6 Area of the property, f_7 Polluted air, f_8 Atractivity of the region for tourists and f_9 Air pollution in result of combustion process.

The results of the models M1 and M2 varies only in the number of interactions used at the rotation. For the output interpretation the model M1 is used (models did not vary much). The first factor "health" explains after the rotation 13.976 % of the overall variances of the set of variables. It is characteristic of the high loadings by the following variables: Number of doctors for 1,000 inhabitants (p_{22}), Number of inhabitants for a doctor (p_{23}), Average number of insured people (p_{26}), Number of chemists and medicine counters (p_{25}), Number of theaters (p_{36}), Density of inhabitants (p_2), Number of galery (p_{35}), Number of hospital (p_{24}), Number of vacancies (p_{20}) and Made investements in environment protection ... (p_{10}). The second factor "leisure activities" explains 9.732 % of the overall variance of the set of variables. The high loadings are reached at the following variables: Number of gymnasiums (p_{40}), Number of playgrounds (p_{39}), Number of stadiums including the indoor ones (p_{41}), Number of cinemas (p_{32}), Number of museums (p_{34}), Number of winter stadiums including the indoor ones (p_{42}), Number of swimming pools and natatoriums (p_{38}), Number of cirques (p_{37}). The third factor "social situation" makes clear 8.545 % of the overall varibality of the set of variables. The high loadings are reached at the following variables: Average old-age monthly pension (p_8), Average monthly income (p_7), Average period of one case of incapacity to work (p_{27}), Average age of the inhabitants (p_3).

The fourth factor "well-being" explaining 7.198 % of the overall variance of the set of variables has the high factor loadings at these variables: Divorce rate (p_{16}), Number of offence (p_{29}), Number of fires (p_{31}), Number of abortions (p_{17}). The fifth factor "density of the inhabitants" makes clear 6.890% of the overall variance of the set of variables. The high factor loadings reach the variables: Overall increase of the inhabitants (p_{18}), Number of accidents (p_{30}), Rate of the registered unemployment (p_{19}), Number of completed flats … (p_5), Number of marriages (p_{15}), Average percentage of the sikness per year (p_{28}). The sixth factor "area of the property" explains 6.575 % of the overall variance of the set of variables. The high loadings are reached at the variables: Length of the roads and motorways (p_{21}), Area of the district altogether (p_1), Number of public libraries including their branches (p_{33}).

The seventh factor "polluted air" explains 5.208 % of the overall variance of the set of variables. The high loadings are characteristic for these variables: Emissions of the basic air polluting substances – sulphur dioxide (p_{12}), Emission of the basic air polluting substances – nitrogen oxide (p_{13}). The last two factors explain less then 5 % of the overall variance of the set of variables. The eighth factor "atractivity of the region for tourists" makes clear 4.996 % of the variance and the ninth factor "air pollution in result of combustion process" makes clear 3.555 % of the variance of the variables. The high loadings at the eighth factor are typical variables Number of collective accommodation establishment (p_6), Number of small proptected areas (p_9), Portion of the selected kinds of properties – farmland from the overall district area (p_4). While at the ninth factor these variables are Emission of the basic substances of the air pollution – carbon monoxide (p_{14}) and Emission of the basic substances of the air pollution – solid emissions (p_{11}).

4. Conclusion

The effect of the modelling of the QL is a nontrivial and complex problem. It is affected by uncertainty. The uncertainty is given by the state of scientific knowledge in this area, a certain degree of error in input data and also by the high degree of the openness of the whole system. The term satisfaction itself is, like the term the QL, very wide and uncertain (multidimensional, complex). Psychological, social, medical a philosophical view projects in delimitation of these terms.

Regional development and the growth in the QL of its citizens belong to the essential goals of regional management. For the regional management, information on citizen satisfaction is an important basis in decision making and selfassessment; that is why it is necessary to assess and measure citizen satisfaction. It is important to identify not only areas in which people fulfil their personal aspirations, but also areas with negative influence on people. Region inhabitants judge their own interests by possibilities and barriers which influence fulfilling their personal needs and interests and unwind their positive or negative relation to the place where they live.

The first task occupied by the modelling of citizen satisfaction with the quality of the environment was processed as the classification task. Its goal was to classify citizens into classes by determining their satisfaction with the quality of the environment. Classification models on the basis of algorithms C5.0, C5.0 with boosting, C&RT and CHAID method were designed. The best results were achieved using algorithm C5.0 and boosting method. On the basis of achieved experience it was corroborated that the algorithm C5.0 is the best in the wide spectrum of classification tasks. The second task was defined clusters – linguistic values 'high', 'middle' and 'low' for QHS, and classify the CR regions based on QHS.

Finally we compare achievement results and apply multidimensional statistical methods to definition new composite indicators. There were two basic methods used, which showed the possibility of obtaining latent variables in the data dealing with the QL in the districts of the CR. By experimentations and application of other methods of FA and by experimentations with PCA it is possible to achieve new results not only in a sociological field. Properly designed models may also serve (Blahuš, 1895): to simplify the description of phenomena in the monitored area, to estimate the indirect measurement of the intermediate and measurable indicators; transform the original variables into a more advantageous form, the creation or verification of a structural theory of the investigated area, etc.

Future work could be focused on research in two levels. It is about the definition of the problem itself, which relates to the problematic of the QL and about the application of the new or hybrid methods. It is mainly the construction of the aggregated indicators for single priority axis SF as well as the suggestion of the aggregated indicators by cross-section of each axis. Another output can be the suggestion of the set of regulations connected to the QL for the regions in the CR and their generalization to the national level, or making a "handbook" for the public administration useful for the process of planning the social policy at the regional level. Final and very actual problem is research of the seniors' QL. Despite the fact, that this problematic is discussed at the international level in Lisbon strategy[2] (Portal Europa, 2011) and at the enquiry of seniors' QL is possible to use e.g. questionnaire WHOQOL-OLD, it is necessary to study possible effects of the systemic changes to the government policy at the national and regional level. The importance of the problematic of the old age in the CR is supported by the demographic and economic indicators. The results of the census in 2001 (CSO, 2011; Vidovičová & Rabušis, 2003) show that in mentioned year there were about 1.5 million people older than 65, which means 14 % of the whole population. By the year 2030 this portion will be 24 % and in the year 2050 up to 33 % of the whole population. In the year 2001 the old-age pension was 6,352 CZK and 140 billion CZK was paid altogether. The difference of gained and invested money in this chapter of the social security was 19 billion CZK (Vidovičová & Rabušis, 2003).

Related to the methods and algorithms, it will be e.g. a using of fuzzy logic and rough sets theory, hybrid fuzzy-rough sets approach, combination of DTs (it means to design a decision forest) etc. Fuzzy logic can be used for an expression of uncertainty into attributes values and rough sets theory can be used for defining "sample" cases into the cases base in Case-based Reasoning algorithm.

5. Acknowledgment

This work was supported by the scientific research project of Ministry of Environment, CR under Grant No: SP/4i2/60/07 with title Indicators for Valuation and Modelling of Interactions among Environment, Economics and Social Relations.

6. References

Albright, S. Ch., Winston, W. L. & Zappe, Ch. (2006). *Data Analysis and Decision Making*, South-Western College Pub, ISBN 0538476125, Mason

---

[2] Lisbon strategy in its social pillar defines people between 55 and 65 as one of the key target groups

Augustinová, M. (2010). *Modelování kvality života pomocí faktorové analýzy* [Quality of life modelling by means of factor analysis], Faculty of Economics and Administration, University of Pardubice, Pardubice (in Czech)

Averill, J. R. & More, T. A. (2000). Happiness, In: *Handbook of Emotions*. Lewis, M., Haviland-Jones, J. M., pp. 663–676, The Guilford Press, ISBN 978-1572305298, New York

Baštecká, B. & Domkařová, P. (2011). Kvalita života, životní styl a vrstevníci v obci [Quality of life, lifestyle and peers in the municipality], 18-09-2011, Available from: <www.kvalitavpraxi.cz/res/data/012/001438.pdf> (in Czech)

Blahuš, P. (1985). *Faktorová analýza a její zobecnění* [Factor Analysis and Its Generalization], SNTL , Praha (in Czech)

Blažej, A. (2005). Kvalita života z aspektu udržateľného rozvoja v 21. storočí [Quality of life from the perspective of sustainable development in the 21st century], *Proceedings Kvalita života a rovnosť príležitostí - z aspektu vzdelávania dospelých a sociálnej práce*, ISBN 80-8068-425, Prešov: Filozofická fakulta Prešovskej univerzity v Prešove, Nov. 2004 (in Czech)

Charles University Environment Centre. (2010). Human Development Index – HDI, 15-09-2011, Available from: <http://www.cozp.cuni.cz/COZP-39.html> (in Czech)

Czech Statistical Office. (2010). Gross Domestic Product, 15-09-2011, Available from: <http://www.czso.cz/csu/redakce.nsf/i/hruby_domaci_produkt(hdp)> (in Czech)

Czech Statistical Office. (2011). Projekce obyvatelstva České republiky [Population projections of the Czech Republic], 23-09-2011, Available from: <http://www.czso.cz/csu/2004edicniplan.nsf/t/B0001D6145/ $File/4025rra.pdf> (in Czech)

Červenka, J. (2011). Jak měřit životní úroveň? [How to measure living standards?], In: *Socioweb 2011*, 15-09-2011, Available from: <http://www.socioweb.cz/ index.php?disp=teorie&shw=114&lst=103> (in Czech)

Diener, E. & Lucas, R. E. (2000). Subjective Emotional Well-being, In: *Handbook of Emotions*. Lewis, M., Haviland-Jones, J. M., pp. 325–337, The Guilford Press, ISBN-13: 978-1572305298, New York

Disman, M. (2005). *Jak se vyrábí sociologická znalost* [How to create sociological knowledge], Karolinum, ISBN 978-80-246-0139-7, Czech Republic (in Czech)

Dragomirecká, E. (2006). Czech Version of the Questionnaire of Quality of Life WHOQOL - translation of the items and scale construction. *Psychiatrie*, Vol 10, No 2, pp. 68 – 73, ISSN 1212-0383

Dragomirecká, E., Bartoňová, J., Motlová, L., Papežová, H., Kožnarová, R. & Šrámková, T. (2006). *Příručka pro uživatele české verze Dotazníků subjektivní kvality života SQUALA* [User guide for the Czech version of the questionnaire of subjective quality of life SQUALA], PCP, ISBN 80-85121-47-6, Prague (in Czech)

Duffková, J. (2006). Životní způsob/styl a jeho variantnost [Life-style and its variability] (Malé zamyšlení nad tím, co všechno se může skrývat pod označením „alternativní životní styl"), *Aktuální problémy životního stylu*, ISBN 80-7308-131-8, Prague: Charles University, 2006 (in Czech)

Dvořáková, Z., Dušková, L. & Svobodová, L. a kol. (2006). *Svět práce a kvalita života: Vliv změn světa práce na kvalitu života* [World of work and quality of life: Impact of changes in the world of work on quality of life], Occupational Safety Research Institute, ISBN 80-86973-08-5, Prague (in Czech)

Halečka, T. (2002). Kvalita života a jej ekologicko-environmentálny rozmer [Quality of life and its ecological and environmental dimension], *Kvalita života a ľudské práva v kontextoch sociálnej práce a vzdelávania dospelých- proceedings*, ISBN 80-8068-088-4, Prešov, Apr. 2001 (in Czech)

Hamplová, D. (2006). Životní spokojenost, štěstí a rodinný stav v 21 evropských zemích [Life satisfaction, happiness and marital status in 21 European countries], In: *Sociologický časopis/Czech Sociological Review*, 16-09-2011, Available from: <http://sreview.soc.cas.cz/uploads/1af3eddc9c2e2c08b0b551f2e43757353f124d35_580_105Hamplova20.pdf> (in Czech)

Keller, J. (1995). *Úvod do sociologie* [Introduction to sociology]. Slon, ISBN 80-85850-06-0, Prague (in Czech)

Kim, J. & Hatfield, E. (2004). Love Types and Subjective Well-being: A Cross-Cultural Study, *Social Behavior and Personality*, Vol 32 (2), pp. 173–182, ISSN 0301-2212

Křivohlavý, J. (2004). Kvalita života vymezení pojmu a jeho aplikace v různých vědních oborech s důrazem na medicínu a zdravotnictví [Quality of life definition and its application in various science disciplines with an emphasis on medicine and health care], *Kvalita života – proceedings*, ISBN 80-86625-20-6, Třeboň (in Czech)

Křivohlavý, J. (2011). *Psychologická pojetí a způsoby zjišťování kvality života* [Psychological concepts and methods of determining the quality of life], 15-09-2011, Available from: <http://www.volny.cz/j.krivohlavy/clanky/c_kvalita.html> (in Czech)

Křupka, J., Kašparová, M., Mandys, J. & Jirava, P. (2009). Quality of Life Investigation Case Study in the Czech Republic, *Sixth International conference on fuzzy systems and knowledge discovery Vol. 1*, ISBN 978-0-7695-3735-1, Tianjin, Aug. 2009

Křupka, J., Jirava, P., Kašparová, M. & Mandys, J. (2010). Approach to Synthesis of Health and Environmental Model, *Selected topics in applied computer science - proceedings*, ISBN 978-960-474-231-8, Iwate, Oct. 2010

Křupka, J., Kašparová, M. & Jirava, P. (2010). Modelování kvality života pomocí rozhodovacích stromů [Quality of life modelling based on decision trees], *E & M Ekonomie a Management*, Vol. 13 (3), pp. 130-146, ISSN 1212-3609 (in Czech)

Křupka, J., Kašparová, M., Jirava, P. & Mandys, J. (2011) Quality of Life Modeling at the Regional Level, In: *Environmental Modeling for Sustainable Regional Development: System Approaches and Advanced Methods*, Olej, V., Obrsalova, I., Krupka, J., pp. 392-415, IGI Global, ISBN 1609601564, USA

Křupka, J., Jirava, P., Mandys, J. & Mezera, F. (2011). Analysis of Selected Regional Quantitative Indicators, *Proc. of the 5th WSEAS International Conferences RES '11, EPESE '11, WWAI '11*, Iasi, July 2011

Křupka, J., Jirava, P., Mandys, J., Mezera, F. & Kašparová, M. (2011). Possibilities of Analysis of Selected Sustainable Development Regional Indicators, *International Journal of Mathematical Models and Methods in Applied Sciences*, Vol. 5 (8), pp. 1372-1379, ISSN 1998-0140

Kubátová, H. (2010). *Sociologie životního způsobu* [Sociology of lifestyle]. Grada Publishing a.s., ISBN 978-80-247-2456-0, Prague (in Czech)

Maimon, O. & Rokach, L. (2005). *Decomposition metodology for knowledge discovery and data mining*, World Scientific Publishing, ISBN 9812560793, London

Maříková, H., Petrusek, M. & Vodáková, A. (1996). *Velký sociologický slovník* [Big sociological dictionary]. Karolinum, ISBN 8071843113, Prague (in Czech)

Ministry of Environment ČR. (2010). Strategický rámec udržitelného rozvoje ČR [Strategic framework for sustainable development in the Czech Republic]. 20.08.2011, Available from:
<http://www.mzp.cz/cz/ strategie_udrzitelneho_rozvoje >

Miovský, M. (2006). *Kvalitativní přístup a metody v psychologickém výzkumu* [Qualitative approach and methods in psychological research.]. Grada Publishing, a.s., ISBN 80-247-1362-4, Prague (in Czech)

Miovská, L. (2011). Dotazník kvality života WHOQOL-BREF a WHOQOL-100 [Quality of life questionnaire WHOQOL-BREF and WHOQOL-100], 20-09-2011, Available from: <http://www.adiktologie.cz/articles/cz/165/904/Dotaznik-kvality-zivota-WHOQOL-BREF-a-WHOQOL-100.html?acc=enb> (in Czech)

Možný, I. (2002). *Česká společnost* [Czech society]. Portál, ISBN 80-7178-624-1, Prague (in Czech)

Műhlpachr, P. (2005). Měření kvality života jako metodologická kategorie [Measuring quality of life as a methodological category], *Proceedings Kvalita života a rovnosť príležitostí - z aspektu vzdelávania dospelých a sociálnej práce*, ISBN 80-8068-425, Prešov: Filozofická fakulta Prešovskej univerzity v Prešove, Nov. 2004 (in Czech)

O'Boyle, C., Browne, J., Hickey, A., Mcgee, H., M. & Joyce, C., R., B. (1993). Schedule for the Evaluation of Individual Quality of Life: a Direct Weighting Procedure for Quality of Life Domains (SEIQoL-DW) - Administration Manual, 16-09-2011, Available from:
<http://epubs.rcsi.ie/cgi/viewcontent.cgi?article=1042&context= psycholrep&sei-redir=1#search=%22Schedule%20Evaluation%20Individual%20Quality%20Life%22>

Oliver, J., P. (1992). The social care directive: Development of a quality of life profile for use in community services for the mentally ill, *Social Work and Social Sciences Review*. Vol. 3(1), pp. 5-45, ISSN 0953-5225

Payne, J. et al. (2005). *Kvalita života a zdraví [Quality of life and health]*, Triton, ISBN 80-7254-657-0, Prague

Phillips, D. (2006). *Quality of Life: Concept, Policy and Practice*, Routledge, ISBN 978-0-415-32355-0, London

Portal Europa. (2011) Evropská unie v České republice. Sociální pilíř lisabonské strategie [The European Union in the Czech Republic], 22-09-2011, Available from: <http://ec.europa.eu/ceskarepublika/abc/policies/ art2377_cs.htm#socpilir> (in Czech)

Půček, M. et al. (2005). *Měření spokojenosti v organizacích veřejné správy – soubor příkladů* [Measurement of satisfaction in public organizations - a set of examples], Ministerstvo vnitra České republiky, ISBN 80-239-6154-3, Prague (in Czech)

Rapley, M. (2003). *Quality of Life Research: A Critical Introduction,* SAGE, Reprint 2007, ISBN 978-0-7619-5456-9, London

Rokach, L. & Maimon, O. (2008). *Data mining with decision trees: Theory and applications,* World Scientific Publishing, ISBN 9812771719, London

Řehulková, O. & Řehulka, E. (2008). Otázky kvality života na základě předchozích výzkumů [Questions of quality of life based on previous research], In: *Kvalita života v souvislostech zdraví a nemoci.* Řehulková, O. et al., pp. 16-30, MSD, ISBN 978-80-7392-073-9, Brno (in Czech)

SPSS Inc. (2007). *Clementine ® 12.0 Modeling Nodes,* ISBN 1-56827-397-5

Stack, S. & Eshleman, J. R. (1998). Marital Status and Happiness: A 17-Nations Study, *Journal of Marriage and the Family, Vol.* 60, Issue 2, pp. 527–537, ISSN 00222445

Šolcová, I. & Kebza, V. (2005). Kvalita života v psychologii: Osobní pohoda (well-being), její determinanty a prediktory [Quality of life in psychology: Personal well-being, its determinants and predictors], *Kvalita života – proceedings,* ISBN 80-86625-20-6, Třeboň, Oct. 2004 (in Czech)

Tokárová, A. (ed.). (2002). *Kvalita života v kontextoch globalizácie a výkonovej spoločnosti.* [Quality of life in the contexts of globalization and performance society] Prešov: Filozofická fakulta Prešovskej univerzity v Prešove, 2002. 199 s. ISBN 80-8068-087-6 (in Czech)

Tokárová, A. (2002). K metodologickým otázkam výskumu a hodnotenia kvality života [The methodological issues of research and evaluation of quality of life], *Kvalita života v kontextoch globalizácie a výkonovej spoločnosti.* ISBN 80-8068-087-6, Prešov, 2002 (in Czech)

Tokárová, A., Kredátus, J. & Frk, F. (ed.). (2005). *Kvalita života a rovnosť príležitostí - z aspektu vzdelávania dospelých a sociálnej práce* [Quality of life and equality opportunities - from the aspect of adult education and social work]. Prešov: Filozofická fakulta Prešovskej univerzity v Prešove, 2005. 908 s. ISBN 80-8068-425-1 (in Czech)

Turban, E., Aronson, J. E. & Liang, T. P. (2005). *Decision Support Systems and Intelligent Systems.* Upper Saddle River, NJ: Pearson Education Inc.

Vaďurová, H. & Műhlpachr, P. (2005). *Kvalita života: teoretická a metodologická východiska* [Quality of life: theoretical and methodological basis]. Masarykova univerzita, ISBN 80 -210 -3754 -7, Brno (in Czech)

Vidovičová, L. & Rabušis, V. (2003). *Senioři a sociální opatření v oblasti stárnutí v pohledu české veřejnosti - zpráva z empirického výzkumu* [Seniors and social provisions in the field of aging in the Czech public view - a report of empirical research], VÚPSV Praha, Brno (in Czech)

Vorlíček, Z. (2008). Strategie Národní politiky kvality v České republice na období let 2008 až 2013 pro vyšší kvalitu života občanů České republiky [Strategy of national policy of quality in the Czech Republic for the period 2008 to 2013 for the higher quality of life of inhabitants in the Czech Republic], *Veřejná správa, čtrnáctideník vlády ČR*, Ministerstvo vnitra, Vol. 11(5), ISSN 1213-6581 (in Czech)

10

Benchmarking Distance Learning Centers with a Multiobjective Data Envelopment Analysis Model

Lidia Angulo Meza, João Carlos Correia Baptista Soares de Mello
and Silvio Figueiredo Gomes Junior
Universidade Federal Fluminense and Universidade Estadual da Zona Oeste
Brazil

1. Introduction

Since the beginning of the 21st century Rio de Janeiro State is concerned with distance learning project. This project is called CEDERJ (for the name in Portuguese). One of CEDERJ's main goal is to contribute with the geographic expansion of undergraduate public education. CEDERJ's expansion in terms of number of local centers and types of courses brings up the need to evaluate CEDERJ globally, since the system consumes public resources.

In this chapter we use advanced models in Data Envelopment Analysis (DEA) to undergo the evaluation of the CEDERJ's centers. We shall notice that since its very beginning DEA has been used in educational evaluation (Charnes et al., 1978), mainly because DEA deals with multiple inputs and outputs and does not need financial figures. Our evaluation is limited to the mathematical undergraduate course because it exists since the very beginning of the CEDERJ.

We first perform a standard DEA evaluation of the CEDERJ centers. Due to the structure of CEDERJ, a problem arises when trying to identify benchmarks for the inefficient centers. In fact, CEDERJ has some centralized decisions and some decisions that are independent for each center. In DEA this means that some variables are not controlled by the centers themselves. Therefore, we introduced a new multiobjective DEA model with non-controllable variables, called MORO-D-ND. This model provides a set of targets for inefficient centers to achieve in order to become efficient. The CEDERJ center may choose among this set the most suitable target. We also introduce a decision rule for this choice. We calculate the Center non radial efficiency index corresponding to each target in the set. To calculate the non radial efficiency we use the efficiency index presented by Gomes Jr et al (2010a).

This chapter is organized as follows. After this introduction we present the CEDERJ in section 2. Basic Data Envelopment Analysis altogether with the MORO-D-ND model and the non radial efficiency index can be found in section 3. Section 4, we present the case study and results. Some final remarks are presented in Section 5.

2. CEDERJ

CEDERJ is the acronym for Rio de Janeiro Center for Distance Learning (in Portuguese *Centro de Educação a Distância do Estado do Rio de Janeiro)*. One of the CEDERJ's main target is to contribute with the geographic expansion of undergraduate public education. This is also one of the targets of public universities in general. A second main target is to grant access to undergraduate education for those who are not able to study in regular hours, usually because of work. Finally, developing the state's high school teachers and offering vacancies in graduate courses are also targets to be achieved.

In june, 2010 the coursed of the CEDERj included Mathematics, Biology, History, Pedagogy, Chemistry, Turism, Physics, Technology in Sciency Computation and Management. CEDERj has 34 (thirty four) centers convering almost all the Rio de Janeiro State, as shown in Figure 1.

Fig. 1. Map of the CEDERj Centers in the Rio de Janeiro State.

In CEDERJ, students have direct contact with tutors, who are of great importance (Soares de Mello, 2003) for they are responsible for helping students with their subjects as well as their motivation. Its pedagogical program is based on advances in the area of information and communication technologies, but also offers practical classes in laboratories. Students receive printed and digital material, which includes videos, animations, interactivity with tutors, teachers, other students and guests. This environment helps creating knowledge.

Its expansion in terms of number of local centers and types of courses brings up the need to evaluate CEDERJ globally, since the system consumes public resources, and also locally, in order to reduce eventual differences.

Gomes Junior *et al* (2008) evaluated CEDERJ courses using the so called elementary multi-criteria evaluation (Condorcet, Copeland and Borda).The authors point out that there is an apparent relation between regions wealth and its position in the final ranking; and a reverse relation between the number of regular universities and the local center's position. In the present study, these variables should be considered when clustering the local centers.

Menezes (2007) made a scientific investigation on distance education, focusing on CEDERJ, analysing how new information and communication technologies impact on time and space organization.

There are many other studies on CEDERJ, yet they are mostly qualitative. Qualitative literature allows different interpretations, and it might become clearer with measurable facts. Our goal is with this quantitative approach to complement the existent qualitative literature, with no intention to replace it.

3. Basic DEA and some new models

DEA classical models, CCR (Charnes *et al.*, 1978) and BCC (Banker *et al.*, 1984), were based on the Farrell (1957) definition for efficiency to evaluate DMUs that use multiple input and multiple outputs. This is done in a radial manner that is reducing inputs or increasing outputs equiproportionally, in which is called radial efficiency. This radial efficiency can be calculated input oriented, reducing all inputs in the same proportion; or output oriented, increasing all outputs in this same proportion.

For inefficient DMU a target is determined, which consist of the inputs or outputs levels the DMU has to attain to become efficient. Models have been developed to present alternative targets. In this section we present the MORO models that provide a set of targets for inefficient DMU.

On the other hand, radial efficiency, suitable in many cases, may not be appropriate for many real cases. As a result, models that deal with different situations have been presented. One situation is relevant for this study: the existence of non discretionary variables. They cannot vary at the discretion of the decision maker. Being part of the analysis they have to be taken into account. Therefore a brief review of these models is also made in this section, pointing out that the main concern of this has been the efficiency evaluation over the targets.

3.1 The MORO models

In many cases the target provided by the DEA classical models may not be viable, due to operational or managerial problems, or simply because we have additional information about the variables. Some alternative models have been presented by Thanassoulis and Dyson (1992) and Zhu (1996). These models provide a target that suits the DMU introducing value judgments about the variables in the model.

One common characteristic is that they only provide one target for inefficient DMUs. If we want a different target a different set of weights must be establish and all the evaluation has to be repeated.

As we have mentioned earlier, the MORO models (Lins et al., 2004, Quariguasi Frota Neto & Angulo-Meza, 2007) provide a set of targets instead of one target for inefficient DMUs. This is possible because of the nature of the multiobjective models, which provide a set of non-dominated solutions (Soares de Mello *et al.*, 2003).

The most common of the MORO models is the MORO-D-CRS that is presented in (1).

This model is very similar to the envelope version of the CCR model. This model allows each variable to change independently, and not in a radial way as the classical DEA models. The ϕ_r factor is the variation for the output r, φ_i factor is the variation for the input i. We have one objective function for each factor, and we try to maximize the factor for the outputs and minimize the factors for the inputs. The restrictions guarantee that we will find

$$Max\ \phi_1$$

$$...$$

$$Max\ \phi_s$$

$$Min\ \varphi_1$$

$$...$$

$$Min\ \varphi_m$$

subject to

$$\sum_{j=1}^{n} y_{rj}\lambda_j = \phi_r y_{ro},\ \text{r=1,...,s}$$

$$\sum_{j=1}^{n} x_{ij}\lambda_j = \varphi_i x_{io},\ \text{i=1,...,m}$$

$$\phi_r \geq 1$$

$$\varphi_i \leq 1$$

$$\phi_r, \varphi_i, \lambda_j \geq 0$$

(1)

projections in the efficient frontier, since the variations of inputs and outputs are independent we replace the inequalities by equalities (Quariguasi Frota Neto & Angulo-Meza, 2007, Lins et al., 2004). The last two restrictions guarantee that the outputs will maintain their levels or increase and the inputs will maintain their levels or decrease. In this way we will obtain targets that dominate the DMU o under evaluation, in an approach similar to the Thanassoulis and Dyson mono-objective model (Thanassoulis & Dyson, 1992). Therefore this model is called MORO with dominance, constant returns to scale.

The last two restrictions can be removed. In this any point in the efficient frontier can be a feasible target for the inefficient DMU, in an approach similar to the Zhu mono-objective model (Zhu, 1996). This model would be called MORO CRS without dominance, or MORO-CRS.

To illustrate these situations, we present Figures 2 and 3 from Soares de Mello et al (2003). In both figures a variable returns to scale frontier is shown. In Figure 2, possible targets for DMU o using the MORO-VRS model are shows. In Figure 3, the MORO-D-VRS is used to determine the targets for DMU o. To obtain the variable returns to scale we introduce the convexity restriction (2) in the model. Such a model would be called MORO-VRS or MORO-D-VRS depending whether we consider dominance or not, as in the aforementioned Figures.

$$\sum_{j=1}^{n} \lambda_j = 1$$

(2)

In contrast to the mono objective models, MORO formulation completely explores the set of projections, ensuring that every projection according to model (1) is found, and determines the maximum number of alternative targets for each DMU. Thus, instead of running several mono-objective problems with different weights, MORO models find the projections through a structured and non-interactive algorithm (Quariguasi Frota Neto & Angulo-Meza, 2007).

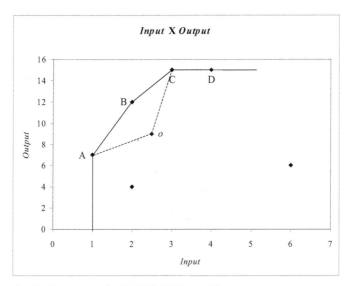

Fig. 2. Targets for DMU *o* using the MORO-VRS model.

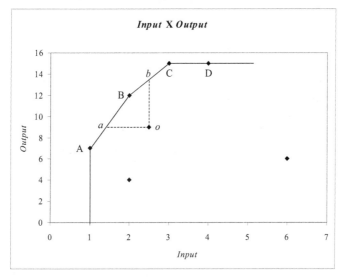

Fig. 3. Targets for DMU *o* using the MORO-D-VRS model.

We have to point out that not only the extreme points are targets for the inefficient DMU but also the linear combinations of these points that lie in the efficient frontier are possible targets. This will happened depending on the method used for solving the multiobjective problem. For example, in Figure 3, for DMU o, the extreme points, targets, are *a*, *B* and *b*, also any point in the segments *aB* and *Bb* are possible targets for DMU *o*. Therefore, in theory we will have an infinite set of targets depending on the method use for solving the multiobjective problem (Soares de Mello *et al.*, 2003).

According to Clímaco et al (2008) the MORO models can be classified in the group that uses multiobjective models to solve problems in DEA.

An efficient DMU is on the Pareto efficient frontier and thus $\phi_r^* = \varphi_i^* = 1$, $\forall\ r,\ i,$ as the equality restrictions of the model require nil value slacks. If this is not the case, the targets for the outputs are given by (3) and the targets for the inputs are given by (4).

$$y_{rj_0}^* = \phi_r^*\ y_{rj_0},\ \forall r \tag{3}$$

$$x_{ij_0}^* = \varphi_i^*\ x_{ij_0},\ \forall i, \tag{4}$$

Therefore, the final value $y_{rj_0}^*$ and $x_{ij_0}^*$ depends on the target chosen by the decision maker and thus we define the values for ϕ_r^* e φ_i^* among the solutions of the MORO model chosen. In this way, alternative targets can be obtained based on the preferences of the decision-maker.

Gomes Junior et al (2010b) stated that the equality restriction on the MORO models may be very restrictive and may present computational problems. They proved that if those restrictions were replaced by inequality constraints, the model will determine the same efficient frontier, that is the same set of targets. The authors called this model with inequalities the MORO-D-R model.

3.2 Non discretionary models in DEA

As mentioned previously, in some real cases DEA classical models do not take into account non discretionary variables. Those variables cannot be modified due to fixed factors of production or external factors. For example: federal employees in Brazil cannot be fired, so they become a fixed of number of resources. This situation can be complicated to deal with in an input oriented DEA model, when there are other inputs that we would like to reduce.

Therefore to deal with non-discretionary variables many researchers have concerned themselves with DEA models for this purpose. The first model was introduced by Banker and Morey (1986) and the input oriented variable returns to scale model is presented in (5).

$$Min\ \theta$$
$$subject\ to$$

$$\sum_{j=1}^{n} y_{rj}\lambda_j \geq y_{rj_0},\ r=1,...,s$$

$$\sum_{j=1}^{n} x_{ij}\lambda_j \leq \theta x_{ij_0},\ i \in D \tag{5}$$

$$\sum_{j=1}^{n} x_{ij}\lambda_j \leq x_{ij_0},\ i \in ND$$

$$\sum_{j=1}^{n} \lambda_j = 1$$

$$\lambda_j \geq 0$$

In the output oriented model (4) we can see that inputs are divided into two groups: controllable and non controllable. We also can see that the maximum equiproportional reduction in all discretionary inputs, maintaining the non discretionary inputs constant. Therefore, the only difference between this model and the standard variable returns to scale DEA model (Banker *et al.*, 1984) is the removal of the factor θ from the right-hand side of the non-discretionary inputs.

They also provided the output oriented variable returns to scale DEA model for non discretionary variables. Analogous to model (1), in the output oriented version, outputs are divided into two groups and then the factor is only multiplied to the controllable outputs.

As pointed out by the authors, the constant returns to scale version of this model can be easily formulated with the exclusion of the convexity constraint (Cooper et al., 2006, Syrjänen, 2004) .Other version was introduced by the same authors Banker and Morey, Camanho et al (2009), Estelle et al (2010) among others.

Golany and Roll (1993) extended Banker and Morey's constant returns to scale model to account for non discretionary variables in both inputs and outputs.

In the Cooper, Seiford and Tone book (2007) a non-controllable model is presented. They stated that restrictions involving non-controllable variables, due to external conditions, should be expressed exactly by equality by a nonnegative combination of the corresponding non-controllable variables. This model is in (6).

$$Min\ \theta$$

$$subject\ to$$

$$\sum_{j=1}^{n} x_{ij}\lambda_j \le \theta x_{ij_0},\ i \in C$$

$$\sum_{j=1}^{n} x_{ij}\lambda_j = x_{ij_0},\ i \in NC \tag{6}$$

$$\sum_{j=1}^{n} y_{rj}\lambda_j \ge y_{rj_0},\ r \in C$$

$$\sum_{j=1}^{n} y_{rj}\lambda_j = y_{rj_0},\ r \in NC$$

$$L \le \sum_{j=1}^{n} \lambda_j \le U$$

$$\lambda_j \ge 0$$

In this model variables are divided in two sets: controllable (C) and non-controllable (NC). The factor h_0 is just for C set. As we can notice the equality restrictions are for the non controllable variables and the inequalities are for the controllable variables. The last restriction imposes an upper bound, U, and a lower bound, L on the sum of λ_j, to take into account the type of returns to scale of the problem (Cooper *et al.*, 2007).

This model represents a different approach when compared to the Banker and Morey model (4), as it takes into account the existence of non-controllable variable in both the input and output sets.

Following these works many other approaches have been proposed. There are alternative models for this variable considering various stages. For example Ruggiero (1996, Ruggiero, 1998), Yang and Paradi (2006).

As Muñiz et al (2006) and Cordero et al (2009) stated, models can be divided regarding the stages needed to perform efficiency analysis, one-stage and multi-stage models. They made a comparison and introduced new models for calculating efficiency with non-discretionary variables.

Camanho et al (2009) classified models for non discretionary factors in two groups: external factors, considering the external conditions where the DMU operates, and internal factors, factors that are internal to the production process but not controlled by the decision makers. They also proposed a method to evaluate efficiency treating the non-discretionary variables according to their classification.

Recently, Estelle et al (2010) presented a new three stage model and made a new comparison of the non discretionary models that uses various stages. Also, Cordero-Ferrera et al (2010) proposed a multi-stage approach based on Tobit regressions. They also used a bootstrap procedure is used to estimate these regressions to avoid potential bias. They illustrated their methodology with an empirical application on Spanish high schools.

All the works above have as main objective to evaluate the efficiency of the DMUs. In the present paper our aim is different; we want to determine a set of targets for the inefficient DMUs, and we do not asses the efficiency of the DMUs in an environment with non-discretionary variables.

3.2.1 Multiobjective models for target determination with non-discretionary variables

As seen previously, the MORO models determine a set of targets for each inefficient DMU. We assume that all variables may change their levels in order to be efficient. In some cases, one target of the set may change the level of one variable at a time. See for example target a for DMU o in Figure 3. If the output in that example is a non-discretionary variable, the decision-maker will choose the target a. Unfortunately, there is no guarantee that the set of targets will always contain a target for any specific non-discretionary variable.

Also, the MORO models allow different degrees of changes in inputs and outputs levels. Thus, to ensure that the set found contains targets that take into account non-discretionary variables we present an extension of the MORO models. The resulting model is in (7) and it is called MORO-D-R-ND, the MORO model with dominance and inequality restrictions with non-discretionary variables, or simply MORO-ND.

In this model we have a factor for every discretionary input (1.. m_i) and output (1.. s_o). We have divided the restrictions of the inputs and outputs in two groups, the one that deal with discretionary variables (D_o *for outputs* and D_i *for inputs*) and the ones that deal with non-discretionary variables (ND_o for outputs and D_i *for inputs*). For the first group, as the variables are allowed to change independently, we have set equalities, in a similar approach as the MORO models (1). For the second one, the variables that we cannot change, we set inequalities similar to the envelope model, in an approach similar to the Banker and Morey model. The last two restrictions of this model we have the dominance restrictions, so for the

$$Max\ \phi_1$$

$$\ldots$$

$$Max\ \phi_{s_o}$$

$$Min\ \varphi_1$$

$$\ldots$$

$$Min\ \varphi_{m_i}$$

$$subject\ to$$

$$\sum_{j=1}^{n} y_{rj}^{D_o} \lambda_j = \phi_r y_{rj_0}^{D_o},\ \forall r \in D_o$$

$$\sum_{j=1}^{n} y_{rj}^{ND_o} \lambda_j \geq y_{rj_0}^{ND_o},\ \forall r \in ND_o \tag{7}$$

$$\sum_{j=1}^{n} x_{ij}^{D_i} \lambda_j = \varphi_i x_{ij_0}^{D_i},\ \forall i \in D_i$$

$$\sum_{j=1}^{n} x_{ij}^{ND_i} \lambda_j \leq x_{ij_0}^{ND_i},\ \forall i \in ND_i$$

$$\phi_r \geq 1,\qquad \forall r \in D_o$$

$$\varphi_i \leq 1,\qquad \forall i \in D_i$$

$$\phi_r, \varphi_i, \lambda_j \geq 0$$

output we can increase or maintain its level and for the input we can reduce or maintain its level.

As the other MORO models, we can obtain a set of targets taking into account the variable that are fixed, for any reason, in the analysis. Obviously, the added advantage is that we do not have to specify an orientation (input or output) for the model, because is a non radial model.

We can also account for the variable returns to scale introducing the convexity restriction (2), and find targets without dominance by eliminating the two last restrictions in model (7).

We can also identify an efficient DMU when $\phi_r^* = \varphi_i^* = 1$, $\forall\ r, i$, as the equality restrictions of the model require nil value slacks. If this is not the case, the targets for the variables are given by equations (3) and (4) in section 2. In this case, the non-discretionary variables will maintain their levels. Once again, the alternative targets can be obtained based on the preferences of the decision-maker.

3.3 DEA non-radial efficiency based on vector properties

It makes no sense to deal with efficiency as a scalar, as this quantity depends on the DMU projection point on the frontier. Thus, the efficiency is characterised by a number and by a direction of projection, characterizing a vector.

Soares de Mello *et al.* (2005) propose an index of vector efficiency. This index has restrictions regarding its utilisation according to the statements of the authors.

In this work we propose the development of a non-radial efficiency index based on the vectorial properties of the problem. These properties define that a DMU must be projected to the efficiency frontier in a direction which is determined by the decision maker, through the choice of the target.

Figure 4 illustrates the concepts which will be used to obtain the vectorial efficiency index. The index was developed for the two dimensional case, as it allows a better visualisation.

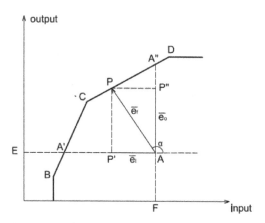

Fig. 4. Basic concepts of vectorial efficiency in a DEA-VRS frontier.

The input or output oriented efficiency in the classic DEA models is calculated by the ratio between the distance from the projection of the DMU on the efficiency frontier to the coordinate axis and the distance between the DMU and the coordinate axis. For DMU A, the input and output oriented efficiencies calculated by the classic DEA models are given, respectively, by the equations (8) and (9).

$$e_{f_i} = \frac{\overline{EA'}}{\overline{EA}} \tag{8}$$

and

$$e_{f_o} = \frac{\overline{FA}}{\overline{FA''}} \tag{9}$$

On the other hand, the complement of the efficiency is the ratio between the distance between the DMU and its projection on the frontier and the distance between the DMU and the coordinate axis. For DMU A, the complements of the input and output oriented efficiency are given, respectively, by the equations (10) and (11).

$$\overline{e_{f_i}} = 1 - e_{f_i} = \frac{\overline{A'A}}{\overline{EA}} \tag{10}$$

and

$$\overline{e_{f_o}} = 1 - e_{f_o} = \frac{\overline{AA''}}{\overline{FA''}} \tag{11}$$

However, we wish to calculate the efficiency index of the DMU when it is projected on the frontier following a non-radial projection.

We supposed that the DMU A is projected on the efficiency frontier on the target determined by the point P. This direction defines an angle α with the horizontal axis. The DMU A has coordinates (x, y). The coordinates of point P are known and denominated (x_E, y_E).

The horizontal projection of point P represents the complement of the efficiency of DMU A if we project only with the input orientation and is given by the equation (12). The vertical projection of point P represents the complement of the efficiency of the DMU A in relation to orientation to output according to the equation (13).

$$\overline{e_{f_i}} = \frac{\overline{AP'}}{\overline{EA}} \tag{12}$$

and

$$\overline{e_{f_o}} = \frac{\overline{AP''}}{\overline{FP''}} \tag{13}$$

In this way, we calculate the complement of the non-radial efficiency of the DMU A when the DMU A has as its projection the target defined by point P by the equation (14).

$$\overline{e_f} = \sqrt{\overline{e_{f_i}}^2 + \overline{e_{f_o}}^2} = \sqrt{\frac{\overline{AP'}^2}{\overline{EA}} + \frac{\overline{AP''}^2}{\overline{FP''}}} \tag{14}$$

When we substitute the coordinate of the points A, E, F, P and their projections P' and P'', we have the equation (15).

$$\overline{e_f} = \sqrt{\left(\frac{x - x_E}{x}\right)^2 + \left(\frac{y_E - y}{y_E}\right)^2} \tag{15}$$

The efficiency index is defined in the interval [0,1] (Cooper & Pastor, 1995). In this way, the efficiency of the DMU A when projected in the target specified by point P is equal to the difference of its complement to the unit and is given by the equation (16).

$$e_f = 1 - \sqrt{\left(\frac{x - x_E}{x}\right)^2 + \left(\frac{y_E - y}{y_E}\right)^2} \tag{16}$$

The DMU target on the frontier is point P. The coordinates of point P are defined by the objective functions of the MORO-D model and calculated by the equations (17) and (18).

$$x_E = \phi.x \tag{17}$$

$$y_E = \varphi.y \tag{18}$$

Substituting the expressions (17) and (18) in the expression (16), we have the non-radial efficiency of a DMU when projected on point P on the frontier, given by the expression (19).

$$e_f = 1 - \sqrt{\left(1-\varphi\right)^2 + \left(1-\frac{1}{\phi}\right)^2} \tag{19}$$

Equation (19) can be generalized for multiple inputs and outputs. This generalization is presented in equation (20).

$$h = 1 - \sqrt{\frac{1}{m}\sum_{i=1}^{m}\left(1-\varphi_i\right)^2 + \frac{1}{s}\sum_{i=1}^{s}\left(1-\frac{1}{\phi_i}\right)^2} \tag{20}$$

The parameters φ_i represents the reductions for each input i, the parameter ϕ_i represents the increase for each output j, for an inefficient DMU to become efficient. The value m and s are the total number of input and outputs, respectively.

3.4 DEA in educational evaluation

DEA has been widely used in educational evaluation. For instance, Abbott & Doucouliagos (2003) measured technical efficiency in the Australian university system. They considered as outputs many variables referring to research and teaching. Abramo et al (2008) evaluated Italian universities, concerning basically scientific production.

The firsts authors went through analysis using various combinations of inputs and outputs, because the choice of the variables can greatly influence how DMUs are ranked, which is similar to what is done the process of variable selection in the present paper. The seconds also verify the importance of choosing the right variables, by comparing the final results with analysis of sensitivity, and observing how different they are.

Abbott & Doucouliagos (2003) introduce the concept of benchmarking as one of DEA strengths, though neither of the articles actually calculates it. Finding benchmarks and anti-benchmarks is important for the study's applicability, since it is the first step to improving the inefficient DMUs. These authors also propose clustering the universities, according to the aspects of tradition and location (urban or not), which in their work, does not significantly affect results.

A more comprehensive review of DEA in education can be found in Soares de Mello et al (2006). More recent works in the subject involve evaluation in secondary education (Davutyan *et al.*, 2010). Sarrico and Rosa (2009) used DEA to evaluate the Portuguese secondary schools. The performance of academic departments was evaluated by Tyagi et al (2009). Tajnikar and Debevec (2008) used DEA to evaluate full-time higher education. In Brazil higher education performance was evaluated by de França et al (2010). A study using DEA and Self-Organizing Maps for CEDERJ centers evaluation was done by Angulo-Meza et al (2011).

4. Case study and results

There are many other studies on CEDERJ, yet they are mostly qualitative or using ordinal methods (Gomes Junior et al., 2008). Qualitative literature allows different interpretations, and it might become clearer with measurable facts. Our goal is with this quantitative approach to complement the existent qualitative literature, with no intention to replace it.

The DMUs being evaluated in the present research are the local centers that offer Mathematics undergraduate course, therefore each of the following variables are related to the Math course in each local center.

The inputs are AI – Number of students enrolled in the course in the first semester of 2005 (this semester was chosen because the course is four years long) and NT – Number of tutors in the first semester of 2009 proxy for the resources used in the center. The output is AF – Number of students that graduated in the first semester of 2009.

There are other professionals, besides tutors, involved in the CEDERJ system, such as those responsible for preparing the material. However, the Math material is the same in every local center, so these professionals should be attributed to each course, not to each local center. Although 24 local centers offer the Math course, only 13 have had graduates in 1/2009, which will be considered in this study.

The centers have no autonomy to formulate the evaluation tests. They are formulated by the central co-ordination. Therefore, the number of students concluding the course is not easy to be manipulated by the centers administrations, and so this number is suitable to be used as an output.

In this evaluation, the non-discretionary variable is the number of students that have enter the course four years before. Obviously, that is because we cannot change the past.

Of course, the number of tutors is a discretionary variable. The number of students that have finished the course is not directly determined by the center. However, it is a consequence of the center educational practices. So this variable can be use a non-discretionary variables (Cordero-Ferrera et al., 2008).

The data set is for the thirteen centers under evaluation is shown in Table 1.

Center	Input 1	Input 2	Output
	AI – 2/2005	NT	AF
Angra dos Reis	60	6	8
Cantagalo	40	7	2
Campo Grande	62	6	1
Itaperuna	36	7	4
Macaé	29	6	3
Paracambi	72	7	9
Petrópolis	79	8	1
Piraí	23	6	6
Saquarema	61	6	2
São Francisco de Itabapoana	20	5	1
São Pedro da Aldeia	62	6	4
Três Rios	60	8	3
Volta Redonda	99	10	10

Table 1. Data set for the educational evaluation.

We use the MORO-D-R-ND model to determine a set of targets for the thirteen centers under evaluation. The TRIMAP (Clímaco & Antunes, 1989) software was used to solve the multiobjective problem. Table 2 shows the results for all the centers.

Center	FACTORS	
	φ	ϕ
Angra dos Reis	1	1
Cantagalo	1	3,85586
	0,214286	1
	0,571429	2,66667
Campo Grande	1	8
	0,125	1
Itaperuna	1	1,87387
	0,428571	1
	0,514286	1,2
Macaé	1	2,10811
	0,375	1
	0,483333	1,28889
Paracambi	1	1,03704
	0,964286	1
Petrópolis	1	10,6126
	0,09375	1
	0,9875	10,5333
Piraí	1	1
Saquarema	1	4
São Fidélis	0,25	1
São Francisco de Itabapoana	1	5,04505
	0,15	1
	0,4	2,66667
São Pedro da Aldeia	1	2
	0,5	1
Três Rios	1	3,1952
	0,28125	1
	0,75	2,66667
Volta Redonda	1	1,32793
	0,75	1
	0,99	1,32

Table 2. Set of targets for the centers.

In this Table 2, we note that there are two efficient centers: Angra dos Reis and Piraí. For each inefficient Center we have a different solutions, leading to different set of targets. In Table 3, we present the target for each inefficient Center and also the vectorial efficiency index corresponding to each target calculated according to equation (20).

Center	Target			Vectorial Efficiency Index
	AI – 2/2005	NT	AF	
Angra dos Reis	60	6,00	8,00	1,0000
Cantagalo	40	7,00	7,71	0,2593
	40	1,50	2,00	0,4444
	40	4,00	5,33	0,3054
Campo Grande	62	6,00	8,00	0,1250
	62	0,75	1,00	0,3813
Itaperuna	36	7,00	7,50	0,5337
	36	3,00	4,00	0,5959
	36	3,60	4,80	0,6182
Macaé	29	6,00	6,32	0,4744
	29	2,25	3,00	0,5581
	29	2,90	3,87	0,5714
Paracambi	72	7,00	9,33	0,9643
	72	6,75	9,00	0,9747
Petrópolis	79	8,00	10,61	0,0942
	79	0,75	1,00	0,3592
	79	7,90	10,53	0,0949
Piraí	23	6,00	6,00	1,0000
Saquarema	61	6,00	8,00	0,2500
São Fidélis	61	1,50	2,00	0,4697
São Francisco de Itabapoana	20	5,00	5,05	0,1982
	20	0,75	1,00	0,3990
	20	2,00	2,67	0,2446
São Pedro da Aldeia	62	6,00	8,00	0,5000
	62	3,00	4,00	0,6464
Três Rios	60	8,00	9,59	0,3130
	60	2,25	3,00	0,4918
	60	6,00	8,00	0,3505
Volta Redonda	99	10,00	13,28	0,7531
	99	7,50	10,00	0,8232
	99	9,90	13,20	0,7575

Table 3. Targets and efficiency indexes for the centers.

Among the set of target, the decision maker can choose one target from the set, according to many managerial needs. In this case, we have chosen the target which provides the higher efficiency index. In Table 3, we have highlighted the target with higher efficiency index for each Center.

In the case where any these targets are not feasible, other multiple objective methods, for example Pareto Race (Korhonen & Wallenius, 1988), can be used to determine an alternative target from the efficient frontier.

5. Conclusion

In this paper, we have performed an educational evaluation of the CEDERJ Centers, using an advance multiobjetive DEA model. The multiobjective model was chosen due to its flexibility in providing alternative targets. Moreover, the vectorial efficiency index provides an effective tool to help the decision maker in choosing the most suitable benchmark for each DMU. The proposed model has proved to be usefull in a managerial and operational aspect.

Comparing our model with classical DEA we notice that both models classify the same DMUs as efficient (it does not change the efficient frontier). However, in our model the inefficient DMUs have the flexibility to achieve a higher efficient index. This characteristic may have decision maker to better accept DEA results as a managerial tool.

Regarding the theorical research, future works involved the comparison between the results obtained with the MORO models and the MORO-R models both with non discretionary variables and dominance constraints.

In relation to the case study, we intend to undergo an evaluation of the other courses, and also involving other CEDERJ Centers.

6. Acknowledgment

To FAPERJ for financial support.

7. References

Abbott, M. & Doucouliagos, C. (2003). The efficiency of Australian universities: A data envelopment analysis. *Economics of Education Review*, Vol. 22, No. 1, pp. 89-97, 0272-7757.

Abramo, G.; D'Angelo, C. A. & Pugini, F. (2008). The measurement of italian universities' research productivity by a non parametric-bibliometric methodology. *Scientometrics*, Vol. 76, No. 2, pp. 225-244, 0138-9130.

Angulo-Meza, L.; Biondi Neto, L.; Brandão, L. C.; Andrade, F. V. S.; Soares de Mello, J. C. C. B. & Coelho, P. H. G. (2011). Modelling with self-organising maps and data envelopment analysis: A case study in educational evaluation. Chapter 4. Pages 71-88. In: Self organizing maps, new achievements, Josphat Igadwa Mwasiagi (Ed.) Intech, Vienna.

Banker, R. D.; Charnes, A. & Cooper, W. W. (1984). Some models for estimating technical scale inefficiencies in data envelopment analysis. *Management Science*, Vol. 30, No. 9, pp. 1078-1092, 0025-1909

Banker, R. D. & Morey, R. C. (1986). Efficiency Analysis for Exogenously Fixed Inputs and Outputs. *Operations Research*, Vol. 34, No. 4, pp. 513-521

Camanho, A. S.; Portela, M. C. & Vaz, C. B. (2009). Efficiency analysis accounting for internal and external non-discretionary factors. *Computers and Operations Research*, Vol. 36, No. 5, pp. 1591-1601

Charnes, A.; Cooper, W. W. & Rhodes, E. (1978). Measuring the efficiency of decision-making units. *European Journal of Operational Research*, Vol. 2, pp. 429-444, 0377-2217.

Clímaco, J. C. N. & Antunes, C. H. (1989). Implementation of an user friendly software package - a guided tour of TRIMAP. *Mathematical and Computer Modelling*, Vol. 12, pp. 1299-1309

Clímaco, J. C. N.; Soares de Mello, J. C. C. B. & Angulo-Meza, L. (2008). Performance Measurement - From DEA to MOLP. IN Adam, F. & Humphreys, P. (Eds.) *Encyclopedia of Decision Making and Decision Support Technologies*. IGI Global, Pennsylvania.

Cooper, W. W. & Pastor, J. T. (1995). Global Efficiency Measurement in DEA. Alicante, Espanha, Departamento de Estadística e Investigación Operativa, Universidad de Alicante.

Cooper, W. W.; Seiford, L. & Tone, K. (2006). *Introduction to data envelopment analysis and its uses: with DEA-solver software and references*, Springer Science, USA.

Cooper, W. W.; Seiford, L. M. & Tone, K. (2007). *A Comprehensive Text with Models, Applications, References and DEA-Solver Software*, Springer, New York.

Cordero-Ferrera, J. M.; Pedraja-Chaparro, F. & Salinas-Jimenez, J. (2008). Measuring efficiency in education: An analysis of different approaches for incorporating non-discretionary inputs. *Applied Economics*, Vol. 40, No. 10, pp. 1323-1339

Cordero-Ferrera, J. M.; Pedraja-Chaparro, F. & Santín-González, D. (2010). Enhancing the inclusion of non-discretionary inputs in DEA. *Journal of the Operational Research Society*, Vol. 61, No. 4, pp. 574-584

Cordero, J. M.; Pedraja, F. & SantÃ-n, D. (2009). Alternative approaches to include exogenous variables in DEA measures: A comparison using Monte Carlo. *Computers and Operations Research*, Vol. 36, No. 10, pp. 2699-2706

Davutyan, N.; Demir, M. & Polat, S. (2010). Assessing the efficiency of Turkish secondary education: Heterogeneity, centralization, and scale diseconomies. *Socio-Economic Planning Sciences*, Vol. 44, No. 1, pp. 35-44

de França, J. M. F.; de Figueiredo, J. N. & dos Santos Lapa, J. (2010). A DEA methodology to evaluate the impact of information asymmetry on the efficiency of not-for-profit organizations with an application to higher education in Brazil. *Annals of Operations Research*, Vol. 173, No. 1, pp. 39-56

Estelle, S. M.; Johnson, A. L. & Ruggiero, J. (2010). Three-stage DEA models for incorporating exogenous inputs. *Computers and Operations Research*, Vol. 37, No. 6, pp. 1087-1090

Farrell, M. J. (1957). The Measurement of Productive Efficiency. *Journal of Royal Statistical Society Series A*, Vol. 120, No. 3, pp. 253-281

Golany, B. & Roll, Y. (1993). Some extensions of techniques to handle non-discretionary factors in data envelopment analysis. *Journal of Productivity Analysis*, Vol. 4, No. 4, pp. 419-432

Gomes Junior, S. F.; Soares de Mello, J. C. C. B. & Angulo-Meza, L. (2010a). Índice de Eficiência Não Radial em DEA baseado em Propriedades Vetoriais. *XLII SBPO - Simpósio Brasileiro de Pesquisa Operacional*, Bento Gonçalves - RS.

Gomes Junior, S. F.; Soares de Mello, J. C. C. B.; Angulo-Meza, L.; Chaves, M. C. d. C. & Pereira, E. R. (2010b). Equivalências em modelos MOLP-DEA que fornecem bechmarks para unidades ineficientes o modelo MORO-D-R. *Revista INGEPRO*, Vol. 2, No. 3, pp. 14-24

Gomes Junior, S. F.; Soares de Mello, J. C. C. B. & Soares de Mello, M. H. C. (2008). Utilização do método de Copeland para avaliação dos pólos regionais do CEDERJ.

Rio's international journal on sciences of industrial and systems engineering and management, Vol. 2, No. 4, pp. 87-98, 1982-6443.

Korhonen, P. & Wallenius, J. (1988). A Pareto Race. *Naval Research Logistics*, Vol. 35, pp. 615-623

Lins, M. P. E.; Angulo-Meza, L. & Moreira da Silva, A. C. (2004). A multi-objective approach to determine alternative targets in data envelopment analysis. *Journal of the Operational Research Society*, Vol. 55, No. 10, pp. 1090–1101

Menezes, E. P. (2007). A espacialidade e a temporalidade da educação a distância: O caso do CEDERJ/CECIERJ. *13º Congresso Internacional de Educação a Distância*, september, 2007, Curitiba.

Muñiz, M.; Paradi, J.; Ruggiero, J. & Yang, Z. (2006). Evaluating alternative DEA models used to control for non-discretionary inputs. *Computers and Operations Research*, Vol. 33, No. 5, pp. 1173-1183

Quariguasi Frota Neto, J. & Angulo-Meza, L. (2007). Alternative targets for data envelopment analysis through multi-objective linear programming: Rio de Janeiro Odontological Public Health System Case Study. *Journal of the Operational Research Society*, Vol. 58, pp. 865–873

Ruggiero, J. (1996). On the measurement of technical efficiency in the public sector. *European Journal of Operational Research*, Vol. 90, No. 3, pp. 553-565

Ruggiero, J. (1998). Non-discretionary inputs in data envelopment analysis. *European Journal of Operational Research*, Vol. 111, No. 3, pp. 461-469

Sarrico, C. S. & Rosa, M. J. (2009). Measuring and comparing the performance of Portuguese secondary schools: A confrontation between metric and practice benchmarking. *International Journal of Productivity and Performance Management*, Vol. 58, No. 8, pp. 767-786

Soares de Mello, J. C. C. B.; Angulo-Meza, L.; Gomes, E. G.; Serapião, B. P. & Lins, M. P. E. (2003). Análise de Envoltória de Dados no estudo da eficiência e dos benchmarks para companhias aéreas brasileiras. *Pesquisa Operacional*, Vol. 23, No. 2, pp. 325-345

Soares de Mello, J. C. C. B.; Gomes, E. G.; Angulo-Meza, L.; Soares de Mello, M. H. C. & Soares de Mello, A. J. R. (2006). Engineering Post-Graduate Programmes: A Quality and Productivity Analysis. *Studies in Educational Evaluation*, Vol. 32, pp. 136-152, 0191-491X.

Soares de Mello, M. H. C. (2003). Uma experiência presencial em EAD: o caso CEDERJ. . *XXVI CNMAC*, september, 2003, São José do Rio Preto.

Syrjänen, M. J. (2004). Non-discretionary and discretionary factors and scale in data envelopment analysis. *European Journal of Operational Research*, Vol. 158, No. 1, pp. 20-33

Tajnikar, M. & Debevec, J. (2008). Funding system of full-time higher education and technical efficiency: Case of the University of Ljubljana. *Education Economics*, Vol. 16, No. 3, pp. 289-303

Thanassoulis, E. & Dyson, R. G. (1992). Estimating preferred target input-output levels using Data Envelopment Analysis. *European Journal of Operational Research*, Vol. 56, No. 1, pp. 80-97

Tyagi, P.; Yadav, S. P. & Singh, S. P. (2009). Relative performance of academic departments using DEA with sensitivity analysis. *Evaluation and Program Planning*, Vol. 32, No. 2, pp. 168-177

Yang, Z. & Paradi, J. C. (2006). Cross firm bank branch benchmarking using "handicapped" data envelopment analysis to adjust for corporate strategic effects *Proceedings of the 39th Hawaii International Conference on System Sciences*, Hawaii.

Zhu, J. (1996). Data envelopment analysis with preference structure. *Journal of the Operational Research Society*, Vol. 47, No. 1, pp. 136-150

Permissions

The contributors of this book come from diverse backgrounds, making this book a truly international effort. This book will bring forth new frontiers with its revolutionizing research information and detailed analysis of the nascent developments around the world.

We would like to thank Yair Holtzman, for lending his expertise to make the book truly unique. He has played a crucial role in the development of this book. Without his invaluable contribution this book wouldn't have been possible. He has made vital efforts to compile up to date information on the varied aspects of this subject to make this book a valuable addition to the collection of many professionals and students.

This book was conceptualized with the vision of imparting up-to-date information and advanced data in this field. To ensure the same, a matchless editorial board was set up. Every individual on the board went through rigorous rounds of assessment to prove their worth. After which they invested a large part of their time researching and compiling the most relevant data for our readers. Conferences and sessions were held from time to time between the editorial board and the contributing authors to present the data in the most comprehensible form. The editorial team has worked tirelessly to provide valuable and valid information to help people across the globe.

Every chapter published in this book has been scrutinized by our experts. Their significance has been extensively debated. The topics covered herein carry significant findings which will fuel the growth of the discipline. They may even be implemented as practical applications or may be referred to as a beginning point for another development. Chapters in this book were first published by InTech; hereby published with permission under the Creative Commons Attribution License or equivalent.

The editorial board has been involved in producing this book since its inception. They have spent rigorous hours researching and exploring the diverse topics which have resulted in the successful publishing of this book. They have passed on their knowledge of decades through this book. To expedite this challenging task, the publisher supported the team at every step. A small team of assistant editors was also appointed to further simplify the editing procedure and attain best results for the readers.

Our editorial team has been hand-picked from every corner of the world. Their multi-ethnicity adds dynamic inputs to the discussions which result in innovative outcomes. These outcomes are then further discussed with the researchers and contributors who give their valuable feedback and opinion regarding the same. The feedback is then collaborated with the researches and they are edited in a comprehensive manner to aid the understanding of the subject.

Apart from the editorial board, the designing team has also invested a significant amount of their time in understanding the subject and creating the most relevant covers. They scrutinized every image to scout for the most suitable representation of the subject and create an appropriate cover for the book.

The publishing team has been involved in this book since its early stages. They were actively engaged in every process, be it collecting the data, connecting with the contributors or procuring relevant information. The team has been an ardent support to the editorial, designing and production team. Their endless efforts to recruit the best for this project, has resulted in the accomplishment of this book. They are a veteran in the field of academics and their pool of knowledge is as vast as their experience in printing. Their expertise and guidance has proved useful at every step. Their uncompromising quality standards have made this book an exceptional effort. Their encouragement from time to time has been an inspiration for everyone.

The publisher and the editorial board hope that this book will prove to be a valuable piece of knowledge for researchers, students, practitioners and scholars across the globe.

List of Contributors

Alessandro Laureani
University of Strathclyde, United Kingdom

Yair Holtzman
Director WTP Advisors, Business Advisory Services Practice Leader, USA

Wai-Ching Poon, Jayantha Rajapakse and Eu-Gene Siew
Monash University Sunway Campus, Malaysia

Kjell Hausken
Faculty of Social Sciences
University of Stavanger, Stavanger, Norway

Fred Wenstøp
BI Norwegian Business School, Norway

Yair Holtzman and Laura Wells
Director WTP Advisors, Business Advisory Services Practice Leader, USA

Eliane Gonçalves Gomes
Brazilian Agricultural Research Corporation - Embrapa

João Carlos C. B. Soares de Mello, Lidia Angulo Meza and Juliana Quintanilha da Silveira
Fluminense Federal University

Luiz Biondi Neto
Rio de Janeiro State University

Urbano Gomes Pinto de Abreu
Embrapa Pantanal, Brazil

João Carlos Correia Baptista Soares de Mello and Nissia Carvalho Rosa Bergiante
Universidade Federal Fluminense, Brazil

Jiří Křupka, Miloslava Kašparová, Jan Mandys and Pavel Jirava
University of Pardubice, Faculty of Economics and Administration, Czech Republic

Lidia Angulo Meza, João Carlos Correia Baptista Soares de Mello and Silvio Figueiredo Gomes Junior
Universidade Federal Fluminense and Universidade Estadual da Zona Oeste, Brazil